Uncertainty-Based Ship Design Optimization

Zuyuan Liu · Xiao Wei · Baiwei Feng ·
Haichao Chang

Uncertainty-Based Ship Design Optimization

Theory and Method

Science Press
Beijing

Springer

Zuyuan Liu
Wuhan University of Technology
Wuhan, Hubei, China

Baiwei Feng
Wuhan University of Technology
Wuhan, Hubei, China

Xiao Wei
China Ship Development and Design Center
Wuhan, Hubei, China

Haichao Chang
Wuhan University of Technology
Wuhan, Hubei, China

ISBN 978-981-97-2972-2 ISBN 978-981-97-2973-9 (eBook)
https://doi.org/10.1007/978-981-97-2973-9

Jointly published with Science Press
The print edition is not for sale in China mainland. Customers from China mainland please order the print book from: Science Press.

This Springer imprint is published by the registered company Springer Nature Singapore Pte Ltd.
The registered company address is: 152 Beach Road, #21-01/04 Gateway East, Singapore 189721, Singapore

Foreword

Advanced design technology is the source of promoting the high-quality innovative development of ships, the premise of forming advanced and reliable ship programs. The book introduces the uncertainty-based optimization design method into the field of ship development, which can promote the development of ship design in the direction of refinement and greatly improve the level of ship design.

In ship design, there are a large number of uncertain factors that affect the design quality, whether it is the complex and changeable external environment, the lack of designers' subjective understanding, or the improvement of the reliability and accuracy of the design target. If these factors are regarded as fixed values, it is likely to lead to the deterioration of ship performance and even the failure of the scheme. For example, the performance boundary of ships in extreme environmental conditions is not supported, and the optimal performance point is not practical in real marine environmental conditions.

The uncertainty of the design factors is objective, and it is considered as a fixed value to grasp the main contradiction and simplify the design process, which is feasible under the condition that the design technical index and its accuracy are not high. To meet the needs for high-quality development of ships and effectively improve the robustness and reliability of ship design, it is necessary to systematically study the uncertain factors in the design process and put forward more advanced ship design theories and methods.

In the fields of civil engineering, machinery, aerospace, and other fields, uncertainty design methods have been deeply studied and applied. In the field of ship structure analysis, a lot of research work has been carried out using uncertainty methods. However, in the field of ship overall design, uncertainty design optimization has not been systematically studied.

Combined with the research results of relevant projects of the National Natural Science Foundation of China and the Ministry of Industry and Information Technology projects, the author of this book has carried out relatively comprehensive basic research on ship uncertainty optimization design methods. They systematically combed through the results to form a book. The book focuses on the basic bottleneck problems existing in ship uncertainty optimization design based on the concept of reliability and robust design, uncertainty classification, uncertainty design modeling, uncertainty analysis transfer method, etc.; reveals the influence mechanism and law of multi-source random and cognitive uncertainty factors on ship design schemes; and forms a mixture optimal design method of ship shape under uncertainty.

The book is innovative. It will provide guidance and reference for further study in this field. Of course, since the book is an exploratory study, there are inevitably imperfections. Attracting the attention of peer experts and promoting the continuous maturation of the research through mutual discussion, I think this is also the original intention of the publication of the book.

January 2023

Yingfu Zhu
Academician of Chinese Academy of Engineering
Beijing, China

Preface

Current ship design is mostly a deterministic design process. However, in the actual development process, the ship design solution will be affected by many uncertainties, which may lead to the ship performance failing to meet the intended design goals, or even failing. Therefore, in order to ensure the robustness and reliability of the ship design, it is necessary to fully consider the influence of various uncertainty factors at the early stage of design, and carry out the research on the ship uncertainty optimization.

Ship uncertainty optimization is a design method to optimize the design space according to the robustness and reliability design requirements under uncertainty. It includes RDO and RBDO, among which, RDO is an optimal design that pursues the stability of system performance and reduces the impact of uncertainty on system performance, and its typical metrics are the mean and standard deviation of the objective function. RBDO focuses on the impact of uncertainty of design variables and parameters on constraints, i.e., the failure probability of constraints.

However, the current ship uncertainty optimization methods are not mature enough, which are mainly manifested in the lack of accurate classification methods for uncertainty factors; serious time consumption of stochastic uncertainty analysis transfer methods; insufficient accuracy and reasonableness of cognitive uncertainty modeling methods; and the lack of analysis transfer methods under mixed stochastic and cognitive uncertainties. This book mainly focuses on the above problems, and carries out the research on classification of the uncertainty, analytical transfer of aleatory uncertainty based on polynomial chaos expansion method, modeling of epistemic uncertainty, mixed uncertainty analysis and propagation method, and finally introduces the application and effectiveness of the method by taking the hull form optimization with uncertainty under mixed uncertainty as an example.

The research work related to this book is supported by the National Natural Science Foundation of China (51979211, 52271327, 52271330) and Green Intelligent Inland Ship Innovation Programme [(2019)358]. We would like to thank the team's teachers and students, as well as the peers in the field of ship design, for their encouragement and support of this book, and especially thank Academician Zhu Yingfu for his guidance of this book.

Limited to the level of the author, there are inevitably some inadequacies in the book, and I sincerely hope that readers will be generous with your advice.

Wuhan, China
February 2024

Zuyuan Liu
Xiao Wei
Baiwei Feng
Haichao Chang

Contents

Chapter 1
Introduction

Currently, most ship designs adhere to deterministic optimisation principles, often neglecting the impact of uncertain factors. This approach frequently results in ship constraints being compromised or failing to meet predetermined performance objectives. Thus, to guarantee the robustness and reliability of ship performance, it is imperative to comprehensively account for the influence of diverse uncertain factors during the initial design phase and undertake investigations into ship uncertainty optimisation. This chapter concentrates on delineating the sources, categorisation, analysis, and propagation of uncertainty, whilst also highlighting the principal challenges encountered in ship uncertainty optimisation.

1.1 Sources and Classification of Uncertainty

In practical engineering contexts, numerous uncertainties arise throughout the product life cycle, spanning from design and manufacturing to operation. Identifying the sources of these uncertainties holds significant importance for conducting optimal uncertainty design. Scholars worldwide in related fields have extensively researched this topic. Wei [1] addressed the uncertainty inherent in practical engineering by examining mathematical modelling and solutions, categorising the sources of uncertainty into three main types: model uncertainty arising from the mathematical modelling of real structures, data uncertainty related to system input parameters (such as measurement data), and computational errors occurring during numerical calculations. Yu [2] classified uncertainty into physical uncertainty directly linked to physical quantities, statistical uncertainty arising from limited sample sizes during statistical inference, and model uncertainty stemming from the data modelling process. Tailoring to specific engineering contexts, Lan [3] categorised uncertainty affecting structural performance into six dimensions, including uncertainty in

Z. Liu et al., *Uncertainty-Based Ship Design Optimization*,
https://doi.org/10.1007/978-981-97-2973-9_1

measurement information, material parameters, loads, geometric parameters, initial and boundary conditions, and the calculation model.

Once the sources of uncertainty factors have been identified, it becomes imperative to undertake classification research. Presently, the predominant method involves categorising uncertainty factors into aleatory and epistemic uncertainty. Aleatory uncertainty, also referred to as objective uncertainty, primarily pertains to the intrinsic uncertainty inherent in the physical system or the environment itself. This form of uncertainty cannot be mitigated by gathering additional data or information but is inherently irreducible. On the other hand, epistemic uncertainty, also termed subjective uncertainty, arises due to the designer's lack of knowledge concerning a particular behaviour during the modelling process. Unlike aleatory uncertainty, epistemic uncertainty can be minimised through the collection of further data or information and is therefore reducible. These classification methods are widely applied across various domains, including decision analysis, risk assessment, policy analysis, scientific computing, modelling, and simulation.

For intricate product systems, the classification of uncertainty necessitates specific study tailored to the unique characteristics of the system. De Laurentis et al. [4] developed a framework for identifying uncertainty types in the integrated design of aerospace vehicles. Drawing on the control model analogy, De Laurentis classified uncertainty into four categories: input uncertainty (characterised by vagueness or poorly defined parameters), operating environment uncertainty (relating to unknown or uncontrollable external disturbances), model parameter uncertainty (involving errors in mathematical models), and measurement uncertainty (stemming from inaccuracies in actual measurements). Walton [5] conducted a comprehensive examination of significant uncertainties throughout the life cycle of space systems, organising them into developmental, operational, and model uncertainties. With simulation-based design increasingly dominating modern engineering approaches, attention has turned to the uncertainties inherent in modelling and simulation. Generally, simulation prediction uncertainties can be categorised as external or internal uncertainties [6–8]. Internal uncertainty is linked to the simulation model, which can further be divided into model structure and model parameter uncertainties. Model structural uncertainty, also known as non-parametric uncertainty [9], primarily arises from inaccurate model assumptions and the inability to accurately depict physical phenomena [10]. Conversely, model parameter uncertainty, also known as parametric uncertainty, largely stems from limited information available to estimate model parameters.

Oberkampf et al. [11, 12] advocated for the use of the term "errors" to distinctly describe identifiable errors occurring at any stage of modelling and simulation. These errors are not attributable to a lack of knowledge but rather to detectable discrepancies that can be identified by the designer through checks, such as programming errors. A research report from the National Aeronautics and Space Administration (NASA) on multidisciplinary design methods for uncertainty in aerospace vehicles, suggests addressing computational uncertainty classification through a complementary categorisation of two types of uncertainty. One type comprises parametric and model-formal uncertainty, while the other is based on the classification proposed by Oberkampf. This book primarily addresses parametric and model-form uncertainty.

1.2 Modelling Uncertainty

Depending on the specific characteristics of the uncertainty and the available information, different approaches are required to appropriately represent and model uncertainty. Numerous methods for uncertainty modelling have been studied worldwide, including probability methods, convex models, evidence theory, interval analysis, and possibility theory. In general, aleatory uncertainty is characterised using probability methods. This approach boasts a long history, solid theoretical foundation, and greater familiarity compared to other theories. However, for epistemic uncertainty, which lacks an exact probability distribution, imprecise probability methods, also known as non-probability methods, are employed [13, 14]. The primary modelling methods for uncertainty are outlined in Fig. 1.1.

1.2.1 Aleatory Uncertainty Modelling

With the growing significance of uncertainty analysis in engineering, the probability method has been introduced, which represents uncertainty as a random variable or a stochastic process (time-dependent). This method involves analysing sample data from the research object to make statistical inferences and parameter estimations for its stochastic distribution. Once the stochastic distribution is determined, various characteristic indices of the random variable, such as the mean value, standard deviation, statistical moments, and joint probability properties, can be defined and calculated using the probability method. When there is sufficient information available to estimate the probability distribution of the variable, it can be treated as a random variable. Initially, it is assumed that the random variable follows a particular distribution (such as Gaussian, Poisson, or lognormal), and then the parameters of the distribution are estimated using adequate data or other available information to accurately fit its probability density function (PDF). The choice of distribution can

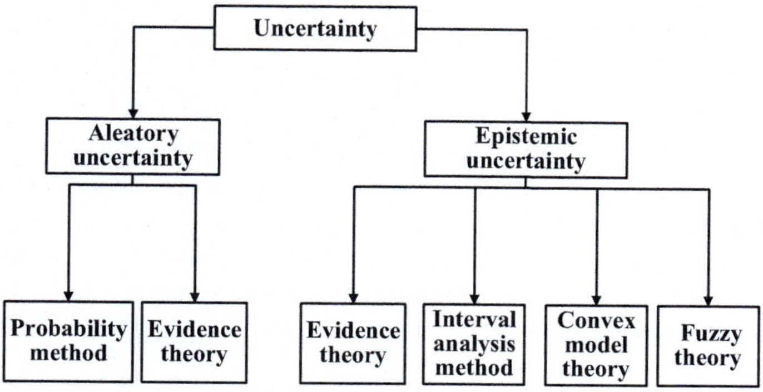

Fig. 1.1 Main modelling methods for uncertainty

be informed by past experience, prior knowledge, expert opinion [15, 16], and the uncertainty characteristics of the variable and its contextual relevance. The parameters of the distribution can be estimated using parametric estimation methods, such as moment estimation or maximum likelihood estimation (MLE) [17]. In cases where the data sample is small, the selection of the statistical distribution model and model fitting can be determined through Bayesian inference of the unbounded Johnson distribution [18].

1.2.2 Epistemic Uncertainty Modelling

While probabilistic methods have been extensively utilised to address uncertainty design problems, they frequently encounter challenges due to insufficient information for constructing probabilistic models. In such scenarios, aleatory uncertainty may degrade into epistemic uncertainty. To manage epistemic uncertainty, research into non-probability methods is gaining traction. Commonly employed methods in non-probability approaches include interval analysis, convex model theory, evidence theory, and fuzzy theory.

1. ***Interval analysis method***

Interval analysis stands out as the most widely employed non-probabilistic method for modelling uncertainty, characterising uncertainty through intervals of variable change. This method is noted for its simplicity and extensive application in practical engineering.

Prior to employing the interval method for analysis, it is imperative to ascertain the variation interval of the uncertainty, thus underscoring the importance of determining the interval range based on available data. Presently, three primary methods are utilised to establish the interval range:

(1) Extreme value method. The minimum value x_{min} and maximum value x_{max} in the data sample constitute the uncertainty interval $[x_{min}, x_{max}]$.
(2) Grey number method. The original data sample is X, and the mean is μ. After the data is arranged from small to large, a monotonically increasing series $X_n^{(0)}$ is obtained (n is the number of samples), and the first k items and (k = 1,2,..., n) of the arranged samples are calculated respectively to get the series $X_n^{(1)}$. Define $\Delta_k = k\mu - X_k^{(1)}$, determine the quantitative evaluation coefficient $s = c\frac{\Delta_{max}}{n}$ of the uncertainty of the data through gray constant c, and determine the interval of the data as $[\mu - 3s,\ \mu + 3s]$.
(3) Statistical method. The mean μ and standard deviation σ of the data sample are calculated and the change interval is determined using a $[\mu - 3s,\ \mu + 3s]$ based on the principle of 3σ.

Donglin [19] conducted a comparison of reliability analysis results obtained using different interval ranges, employing the extreme value, grey number, and statistical methods. The findings indicate that the extreme value method tends to overestimate the product reliability, while the reliability derived from the statistical method proves to be the most conservative. The grey number method yields results falling between these two extremes. In cases of abnormal data, the interval obtained through the grey number method is narrower than that derived from the statistical method. However, upon the removal of abnormal data, the intervals obtained from both methods become nearly identical. Additionally, the value of the grey constant in the grey number method is deliberated upon. Typically, a value of $c = 2.5$ is adopted, although a smaller value may be utilised when data collection is precise.

In the interval analysis method, addressing uncertainty entails grappling with a range of technical challenges, given that uncertainty is expressed as an interval range rather than a precise numerical value. These challenges manifest in the formulation of mathematical transformation models and the resolution of optimisation problems using the interval method. Chao [20] introduced a framework for interval-based uncertainty optimisation theory and algorithmic systems. Within this framework, two types of nonlinear interval number optimisation transformation models were proposed, drawing on mathematical programming theory. These models, termed the interval order relation transformation model and the interval possibility transformation model, aim to convert uncertainty optimisation problems into deterministic optimisation problems.

The interval order relation serves as a criterion for assessing the quality of an interval based on the designer's preference for either the central $f^c(X)$, radius $f^\omega(X)$, or boundary of the interval. Leveraging this order relation, the interval order transformation model converts the uncertain objective function into a deterministic multi-objective optimisation problem. The resulting converted deterministic optimisation model is:

$$\begin{aligned} &\min\,(f^c(X), f^\omega(X)) \\ &\text{s.t.}\quad P(M_i^I \leqslant N_i^I) \geqslant \lambda_i \end{aligned}$$

In the interval possibility transformation model, the uncertain objective function undergoes conversion into a deterministic objective function by incorporating a performance interval V^I. This formulation aims to maximise the probability associated with the objective function $\mathrm{P}(f^I(\mathrm{X}) \leq V^I)$. Consequently, the resulting transformed deterministic optimisation model is:

$$\begin{aligned} &\min\,(f^c(X), f^\omega(X)) \\ &\text{s.t.}\quad P(M_i^I \leqslant N_i^I) \geqslant \lambda_i \end{aligned}$$

When comparing the two transformation models, it is observed that both involve transforming constraints based on interval possibility. However, the interval possibility transformation model extends this transformation to include the objective function, resulting in a single-objective optimisation problem. This model offers a unified and straightforward mathematical formulation. Nonetheless, its effectiveness is contingent upon decision-makers possessing a profound understanding of the optimisation problem to determine the performance interval. Consequently, its applicability may be inferior to that of the interval ordered relational transformation model.

During the operational process of interval numbers, traditional interval algorithms typically conduct mathematical operations using the interval boundaries. However, the interval method, functioning as a conservative estimation analysis technique, tends to expand the interval range of the result during interval operations, resulting in interval expansion and potentially diminishing the practical significance of the outcome. Research indicates that the frequency of occurrences during the process correlates directly with the extent of interval expansion for the same parameter. Moreover, stronger correlations between different parameters lead to greater interval expansion. In efforts to address this issue, Lan [3] analysed various methods including the interval truncation method, changing the order of operations, and subinterval perturbation. Ultimately, Lan employed the affine algorithm to restrict interval expansion, while also discussing its limitations. Zhou [21], in conjunction with Taylor series, ensured each interval parameter appeared only once, thereby reducing interval expansion. Ling [22] introduced the pan-gray method for interval analysis to counteract conservative calculation results associated with interval analysis applications. A comparative analysis of the pan-gray operation rules and interval operation rules reveals identical addition operation rules. However, for subtraction, multiplication, and division operations, the lower and upper bounds of the solution interval obtained using pan-gray operation rules are calculated based on the lower and upper bounds of the pan-gray numbers involved in the operation, respectively. In contrast, the solution interval's lower and upper bounds calculated using interval numbers are determined by the limit values of mixed operations involving the lower and upper bounds of the interval numbers included in the operation. Consequently, the solution interval calculated using pan-gray numbers tends to be smaller than the solution interval calculated using interval numbers, thereby mitigating interval expansion.

2. ***Convex model theory***

The convex model, introduced by Ben-Haim et al. [23] in 1990, presents a more comprehensive approach utilising convex sets to depict uncertainty. This model encompasses various formulations such as the energy constraint model, interval model, ellipsoid model, envelope constraint model, constraint model, and Fourier constraint model. A notable example within the convex model framework is the ellipsoidal model, defined by $\mathrm{X^TWX} \leq \mathrm{a}$, where X represents an uncertain parameter vector, W denotes a positive definite matrix, and a signifies a positive real number. Under this formulation, the uncertainty represented by X is conceptualised as an ellipsoid, as opposed to a hypercube defined by lower and upper boundaries. This deviation is rationalised by recognising that uncertainties are not always independent of each other and do not consistently reach the boundary simultaneously. Consequently, in practical applications, it is common to employ convex models to illustrate correlations among uncertainties.

In cases involving 2- and 3-dimensional uncertainties, the aforementioned convex models manifest as 2-D and 3-D ellipsoids, respectively. Constructing a high-dimensional ellipsoid model necessitates the support of optimisation algorithms to effectively manage the number of samples. Given a specific data sample, an ellipsoid can encompass all sample points if it is sufficiently large. Consequently, numerous ellipsoid models can be constructed, with the smallest ellipsoid representing the most rational uncertainty space, aiming to minimise the uncertainty space. Thus, constructing a high-dimensional ellipsoid model entails resolving the optimisation problem of determining the positive definite matrix W with the smallest ellipsoid volume. Gui [24], leveraging the fundamental principles of nonlinear programming, transformed the minimum enclosing ellipsoid problem into a convex optimisation problem. Subsequently, a stable and efficient algorithm is devised for constructing an ellipsoid model. Furthermore, through comprehensive analysis of the morphological characteristics of interval and ellipsoidal models, fundamental concepts such as expectation, variance, covariance, and total matrix are introduced based on the ellipsoidal model. The covariance matrix between uncertain variables is proven to be the inverse matrix of the characteristic matrix of the ellipsoidal model.

In addition to the ellipsoidal model, several convex models of alternative geometric shapes have found application in practical engineering. Pan et al. [25] introduced an enhanced convex model utilising a parallelepiped, which establishes a 2-dimensional model based on the edge intervals and associated angles of parameters within its uncertainty domain. By mapping higher dimensions onto a 2-dimensional plane and constructing a parallelogram model, they subsequently convert the parallelepiped model into an interval model in affine space through affine coordinate transformation. Ling [26] proposed a novel uncertainty modelling approach based on principal component analysis. This method involves deriving orthogonal eigenvector directions through principal component analysis, transforming uncertain parameter data to the characteristic directions via linear transformation, and obtaining intervals within the new coordinate space. The intersection domain of the traditional interval model and the transformed interval model is utilised as the uncertain convex set model, ensuring a more comprehensive and generalised construction of the convex model, as opposed to one specific to a particular geometric shape.

3. ***Evidence theory***

The evidence theory, proposed by Dempster in 1967 and further refined by his student Shafer in 1976, is also known as Dempster-Shafer evidence theory (D-S evidence theory). It quantifies uncertainty based on known evidence regarding a proposition to ascertain credibility and likelihood. Utilising the interval analysis method, evidence theory subdivides intervals into units known as focal elements, each of which is assigned a basic probability assignment (BPA). For a specific event, the belief measure (Bel), denoted by $\mathrm{Bel(A)} = \sum_{B|B\subseteq A} m(B)$, represents the lower bound of the probability of evidence definition. This measure is computed as the sum of BPAs assigned to all focal elements completely contained within the event. Conversely, the plausibility measure (Pl), given by $\mathrm{PI(A)} = \sum_{B|B\cap A\neq\phi} m(B)$, represents the upper limit. Pl signifies the sum of basic plausibility numbers associated with evidence fully supporting proposition A. Moreover, PI denotes the sum of basic confidence numbers that either fully or partially support proposition A. Together, these measures establish a probability interval that characterises the uncertainty surrounding statement A.

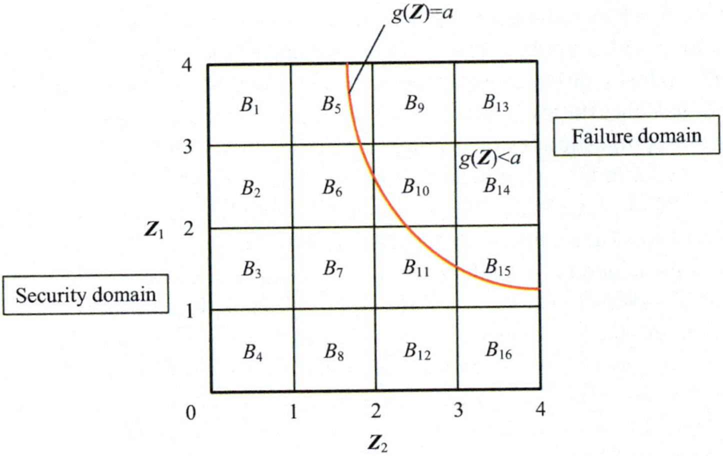

Fig. 1.2 The joint evidence space of the interval variables

To exemplify the 2-dimensional scenario, let us consider two interval variables, Z_1, and Z_2. The joint evidence space of these interval variables is depicted in Fig. 1.2.

The focal elements fully contained within the failure domain A are B_9, B_{13}, and B_{14}. In addition to B_9, B_{13}, and B_{14}, other focal elements intersecting the failure domain A comprise B_5, B_6, B_{10}, and B_{11}, B_{15}. Thus, the total number of focal elements is:

$$\begin{cases} \mathrm{Bel}(A) = m(B_9) + m(B_{13}) + m(B_{14}) \\ \mathrm{Pl}(A) = m(B_9) + m(B_{13}) + m(B_{14}) + m(B_5) + m(B_6) + m(B_{10}) \\ \qquad\quad + m(B_{11}) + m(B_{15}) \end{cases}$$

Information or evidence used to gauge confidence and likelihood can originate from various sources, including experimental data, theoretical evidence, and expert opinions regarding parameter reliability or event occurrences. These pieces of evidence can then update probability measures through specific combinatorial rules. While this method exhibits limitations in handling inconsistent data sources, which might result in unreliable combinations of evidence, numerous scholars have endeavoured to enhance evidence theory and have applied it to tasks such as risk assessment, decision-making, and optimal design.

4. ***Fuzzy theory***

Fuzzy theory, an extension of fuzzy sets and fuzzy logic theory introduced by Zadeh, serves as a tool for modelling uncertainty, particularly in situations with limited information or sparse data [27]. Unlike traditional sets with fixed boundaries, fuzzy sets exhibit flexibility. Let $\widetilde{A}$ denote a fuzzy set, where the membership of a single point x within $\widetilde{A}$, also known as the characteristic function, is expressed by $\mu_{\widetilde{A}}(x)$. Contrary to classical sets, the membership of a point in a fuzzy set $\widetilde{A}$ can vary between 0 and 1; classical sets, in contrast, mandate a membership of either 0 or 1.

Consequently, clear sets can be considered a special case of fuzzy sets. Within fuzzy theory, the membership function expands to depict the degree to which the fuzzy distribution signifies the likelihood of an event occurring.

Both aleatory and epistemic uncertainties can be effectively represented using this theory [28]. Fuzzy theory tends to adopt a more conservative approach regarding confidence levels compared to probability theory. Probability theory proves more advantageous when ample information regarding uncertainty or an accurate prediction model is available [29, 30]. Fuzzy sets and fuzzy theory have also found application in engineering optimisation design and decision-making [31, 32]. By integrating a unified algorithm [33–35] that simultaneously considers both types of uncertainties, fuzzy theory can be effectively combined with probability theory.

1.3 Uncertainty Analysis

1.3.1 Aleatory Uncertainty Analysis

Uncertainty analysis primarily examines the impact of model input uncertainty and model uncertainty on system output, with a focus on assessing and simulating the random characteristics of system output. Typically, uncertainty analysis methods fall into two categories: invasive and non-invasive [36].

Invasive methods primarily entail reconstructing the governing equations and modifying the simulation code to directly incorporate uncertainty into the system. A typical example of this approach is the polynomial chaos expansion (PCE) method. Here, uncertainty is expressed as a random process through orthogonal polynomial expansion, allowing for the analysis of coupled systems with deterministic equations by editing existing analysis code. In contrast, non-invasive methods do not require modification of the simulation code, rendering them more convenient. The PCE method can also address uncertainty problems in a non-invasive manner [37, 38].

The decision to employ an intrusive approach depends on the specific problem and subject matter. For instance, in the analysis of uncertain conditions in engineering structures, the invasive method entails randomly expanding the stiffness matrix. Subsequently, the uncertainty of system responses such as displacement and mode is analysed, necessitating modifications to the finite element code.

This chapter primarily presents widely used non-invasive methods, which include the Monte Carlo (MC) method, the PCE method, and specific methods for reliability analysis.

1. ***MC method***

The MC method belongs to a class of algorithms [39, 40] that conduct repetitive sampling and simulation to acquire statistics for the response. This method is conceptually straightforward and easily implementable as a statistical sampling technique.

In this approach, the input random parameters are sampled repeatedly and independently from a given distribution. For each sample of the input, the output response is computed, and desired statistical properties such as statistical moments and PDFs are estimated. With a sufficient number of samples, the MC method can furnish results of statistical analysis with arbitrary precision. Consequently, it is often utilised as a benchmark to assess the strengths and weaknesses of uncertainty analysis techniques.

The accuracy of the MC method is intricately linked to the number of samples, with a tenfold increase in accuracy necessitating a 100-fold increase in sample size. To address this challenge, various enhanced MC methods based on different sampling techniques have been developed, demonstrating greater efficacy compared to random sampling approaches. Among these techniques, importance sampling, also known as weighted sampling [41], stands out. This method mitigates the error associated with the established PDF through importance sampling [42, 43]. Cao et al. [44] have additionally proposed leveraging the first-order sensitivity information of the target response to expedite the convergence of the MC method. This sensitivity enhancement approach can enhance the accuracy significantly, especially when estimating the failure probability of small orders of magnitude.

2. ***PCE method***

The MC method is frequently employed in engineering applications; however, it necessitates a large number of samples to achieve convergence. Utilising the MC method directly for optimisation design, which entails numerous simulation calculations, is evidently not feasible.

Compared with the commonly used MC method, the PCE method can quantify the random properties of outputs under the joint action of multiple uncertain parameters with lower computational costs. It offers more advantages than the MC method in uncertainty optimisation design. Proposed by Wiener in 1938, this method provides an efficient quantification approach for uncertainty analysis, particularly in aleatory analysis. The PCE method can be viewed as the optimal mathematical approach for constructing and obtaining the model response surface containing uncertain parameters in the form of high-dimensional polynomials. Augustin et al. [45] discuss the advantages and limitations of the PCE method and related expansion techniques Oladyshkin et al. [46] demonstrate the application of PCE for robust design under the influence of uncertainty. Sensitivity analysis based on the PCE method [47, 48] has garnered increasing attention, with Sudret et al. [49, 50] illustrating how the classical PCE method can furnish the necessary information for global sensitivity analysis at a reduced computational cost.

The crucial step in the PCE method involves identifying PCE parameters, which can be broadly classified into invasive and non-invasive methods. Invasive methods necessitate the manipulation of governing equations and can occasionally yield semi-analytic solutions for aleatory analysis, with the aleatory Galerkin technique being the most renowned. Originating in structural mechanics [51], this technique has found application in uncertainty modelling of flow field problems [52, 53]. However, this process is notably intricate and burdensome due to the requisite modifications and adjustments to the original model. Consequently, methods such as sparse orthogonality [54] and probabilistic matching points [55, 56] have garnered increasing attention.

The original PCE method relied on Hermite polynomial expansion, which is most suitable for random variables following a normal distribution. However, natural phenomena and uncertainties in engineering typically exhibit more complexity. The parameters of a physical model often do not conform uniformly to a normal distribution but can be appropriately transformed using methods such as Gaussian transform, normal fraction transform [57], or approximate parameter transform [58]. Yet, expanding transformed variables may not always be the optimal choice, potentially resulting in slow convergence [59]. Consequently, Wiener proposed the Gaussian stochastic process, represented by multidimensional Hermite orthogonal polynomials, leading Ghanem et al. x'60] to develop the spectral random finite element method. This approach has found extensive application across disciplines, including structural mechanics and fluid mechanics [60, 61]. Xiu and Karniadakis [59] further extended this concept by suggesting the representation of stochastic processes using the Askey family expansion of orthogonal polynomials. This method serves as a generalisation of the Wiener Hermite chaotic expansion, employing different subsets of Askey-like polynomials and various polynomials for random variables with different distributions. Orthogonal polynomials based on the Askey family [62] extend the PCE technique to generalised polynomial chaos (gPC). Unlike traditional PCE methods tailored to Hermite polynomials and normal distributions, gPC accommodates common probability distributions like gamma, beta, and uniform distributions, offering more versatility. As the scope of applications broadens, the PCE method must adapt to a wider range of distributions. Scholars have proposed a multi-element generalised polynomial chaos (ME-gPC) method [56, 63], which involves decomposing a random space into local elements and implementing gPC within each element. This adaptive segmentation method addresses distribution discontinuities or model response variations, providing a flexible tool for aleatory modelling. However, the ME-gPC method interprets data as precisely known probability distributions, significantly increasing computational demands for multidimensional aleatory problems [64].

3. ***Reliability analysis of aleatory uncertainty***

Reliability analysis involves assessing the uncertainties inherent in a system and its operational environment. In probabilistic reliability analysis, the failure probability *pf* with a constraint g(x) $\leq$ 0 is determined. Typically, the failure probability can be calculated by integration, as follows:

$$p_f = \int_D p(x)\,\mathrm{d}x$$

The failure domain D is defined as $(x) > 0$. The reliability R of the system is given by $R = 1\text{-}p_f$. Such integrals are often challenging to calculate because the joint probability distribution function and the failure domain D are rarely explicitly defined, and multidimensional integrals may be impossible to perform, particularly for complex systems that require extensive analysis.

Therefore, a variety of approximation methods are required, including numerical integration methods, as well as specific alternative integral approximation methods based on reliability analysis. Examples of these methods include the asymptotic approximation based on Laplace's multidimensional integral method [65], the integral-based failure principal domain covering method [66], the fast Fourier transform method [67, 68], the tail modelling method [69, 70], dimensionality reduction methods [71–73], and the first-order reliability method (FORM) and the second-order reliability method (SORM). Among these approximate methods, FORM and SORM are the most widely used in engineering problems [74]. Additionally, several improvement methods have been proposed to enhance computational efficiency [75, 76]. To improve calculation accuracy, the higher-order reliability method (HORM) [77], the second-order matrix method, the third-order matrix method [78], and the high-order matrix method [79] have been introduced. Furthermore, there are FORM-based methods for the system output cumulative distribution function [80], as well as comprehensive studies on the accuracy and scope of application of FORM and SORM [81]. Breitung [82] analysed the asymptotic behaviour of FORM and SORM in detail, with the results indicating that only SORM provides the asymptotic approximation of the integral, whereas FORM may generate uncontrollable phase errors. Asymptotic approximations can also be obtained by FORM if a generalised reliability index is required [83], and it is more commonly employed in engineering applications due to its computational efficiency advantages.

In addition to the above approximation methods, the sampling method can also be used to directly solve the integral of the failure probability. However, given the required confidence level and percentage error requirements, when p_f is very small, the number of samples will be very large and lead to an unbearable computational burden. To address this issue, scholars have developed several effective sampling techniques, such as importance sampling [84–87] and its variants like Adaptive sampling [85], subset module [86], line sampling [87–89], and direction simulation [88–91]. Simultaneously, to further enhance the efficiency of reliability analysis, the response surface method (RSM) can be utilised to replace the exact function of computation time, thereby reducing the computational burden [89–92].

1.3.2 Epistemic Uncertainty Analysis

Epistemic uncertainty analysis can be classified into two categories: one involves robust analysis while the other focuses on reliability analysis:

1. ***Robustness analysis methods considering epistemic uncertainty***

In the presence of aleatory uncertainty, the typical measure of robustness is the standard deviation of the objective function's response value. However, when considering epistemic uncertainty, the interval radius of the objective function's response value is typically adopted as the measure of robustness due to the lack of precise information regarding uncertainty distribution.

2. ***Reliability analysis methods considering epistemic uncertainty***

In engineering, acquiring accurate uncertainty information can be challenging, thereby limiting the application of probabilistic reliability theory. Frequently, while it may be challenging to ascertain the distribution information of uncertainty, determining its bounds can be relatively straightforward. Against this backdrop, numerous scholars have introduced the theory of non-probabilistic reliability. Ben-Haim [90] proposed the notion of non-probabilistic reliability based on convex set theory, suggesting that a system is deemed reliable if it can accommodate uncertainty within a specified range.

Compared with traditional reliability analysis methods, non-probabilistic reliability theory exhibits imperfections, notably in the lack of unified standards for defining reliability indices. Shuxiang et al. [91] proposed a non-probabilistic reliability measurement system and analysis method based on interval analysis. This approach considers the shortest distance from the origin of coordinates to the failure plane in the expanded space of standardised interval variables as the non-probabilistic reliability index, providing explanations from both physical and geometric perspectives. Minrong [92] established an expression for interval reliability, defining the index as the tangent point of the hypercube box and the failure plane. He elucidated the geometric significance of the non-probabilistic reliability index under the interval model. Xinzhou [21] introduced a structural non-probabilistic reliability model based on ellipsoidal convex sets. This model measures structural non-probabilistic reliability using the ratio of the volume of the structural safety domain to the volume of the basic variable domain. Tao [93] proposed an analytical method and a one-dimensional optimisation algorithm for the non-probabilistic reliability index of uniformly bounded convex sets models. He incorporated affine operations into the calculation of the non-probabilistic reliability index for uniformly bounded convex sets models.

In summary, non-probabilistic reliability theory serves as a valuable supplement to traditional probabilistic reliability methods, but its theory, algorithms, and applications require further improvement.

1.4 Optimal Design of Uncertainty

The theory and application of uncertain optimal design constitute one of the most significant challenges in optimisation. The stochastic linear programming method emerged in the mid-twentieth century to address optimisation problems involving parametric randomness [94]. To confront the complexities posed by discrete integer and nonlinear characteristics during the optimisation process, various novel approaches have been introduced. These include stochastic integer programming [95, 96], stochastic nonlinear programming [97, 98], robust stochastic programming [99, 100], and stochastic dynamic programming [101, 102]. These methods are often integrated under the umbrella term "stochastic programming" [103, 104]. In addressing epistemic uncertainty in optimisation, the development of fuzzy programming and hybrid methods combining random and fuzzy techniques has been rapid [105–108].

In optimisation design, two primary approaches are typically employed to address uncertainty factors. Firstly, efforts may focus on reducing the uncertainty inherent in these factors themselves, aiming to either eliminate or minimise the extent of uncertainty. This can be achieved through various means such as process enhancement, the use of high-performance materials, enhancing precision, or reducing tolerances. Secondly, strategies may be directed towards mitigating the impact of uncertainty on product performance or containing it within an acceptable range. This ensures that the product maintains stable and reliable performance even in the face of uncertainty. The former approach, while desirable, is often challenging to achieve and may entail significant costs. Therefore, the latter approach is often pursued, which involves two types of uncertain optimisation design methods:

(1) Robust design optimisation (RDO) aims to enhance product robustness by reducing sensitivity to changes in design parameters, thereby ensuring performance stability.
(2) Reliability-based design optimisation (RBDO) seeks to enhance product reliability and minimise the likelihood of performance failure, particularly under extreme conditions, to ensure the system remains operational under normal circumstances.

1.4.1 Robust Design Optimisation

RDO was initially developed by Taguchi, who introduced the Taguchi method to enhance product quality and render product performance less sensitive to variations in variables beyond the designer's control. In Fig. 1.3, point 1 represents the primary advantage of deterministic optimisation, while point 2 illustrates the primary advantage of robust optimisation. Without considering the influence of uncertain factors, point 1 would offer the smallest objective function value, thus representing the optimal solution. However, when the design variable x fluctuates within the range of $\pm\,\Delta x$, the fluctuation of the objective function ΔF_1 at point 1 exceeds the constraint condition, resulting in an unacceptable product. Conversely, at point 2, the fluctuation change of the objective function ΔF_2 is significantly less than ΔF_1, ensuring that all target responses remain within the constraints despite fluctuations in the design variable. Therefore, the robust optimal solution represents a more stable optimal solution.

RDO should not only aim to optimise target performance but also reduce its sensitivity to uncertainty. Therefore, in the presence of aleatory uncertainty, its

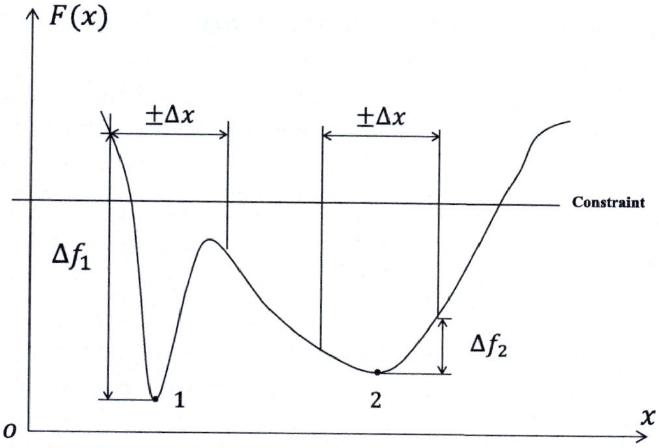

Fig. 1.3 Diagram of deterministic optimal solutions and robust optimal solutions

typical metrics include the mean and standard deviation of the objective function, as expressed by the formula:

$$\begin{aligned} &\min \mu(F) \quad \text{or/and} \quad \sigma(F) \\ &\text{s.t.} \quad g_i \leqslant 0, \quad i = 1, 2, \cdots, n \\ &h_j = 0, \quad j = 1, 2, \cdots, m \end{aligned}$$

where μ(F) and σ(F) are the mean and standard deviation of the objective function, respectively; g_i is the ith inequality constraint; and h_j is the hth equation constraint. RDO has been extensively researched in mechanical, vehicle engineering, aerospace, and other fields. For instance, Daisheng et al. [106] applied the Taguchi method to robustly optimise the mount system using SNR as the fitness index.

Wang [107] conducted a comparison and analysis between the Taguchi method and the Quality Engineering 6-sigma method concerning vehicle smoothness and stability indexes. It was observed that the mean and mean square deviation of the 6-sigma method were smaller. Subsequently, a multi-objective optimisation model for the smoothness and stability of vehicle operation was developed using the 6-sigma method. Guangyong [108] integrated the 6-sigma concept with the RSM to formulate a multi-objective RDO approach. This method was applied to optimise the design of automobile collision and plate forming processes, aimed at enhancing their robustness. Liu [109] devised a non-probabilistic robust optimisation model with the objective of minimising the centre and radius of the interval. This was done to reduce the scrap rate by utilising the interval analysis method, particularly in scenarios with limited uncertain information.

1.4.2 Optimal Design Based on Reliability

At the heart of RBDO lies the management of uncertainty's impact on constraints. Its mathematical formulation under the probabilistic form of constraint is represented as follows:

$$\begin{aligned} &\min F \\ \text{s.t.}\quad &P(g_i \leqslant 0) \geqslant r, \quad i = 1, 2, \cdots, n \\ &\mu(h_j) = 0, \quad j = 1, 2, \cdots, m \end{aligned}$$

where $P(g_i \leqslant 0)$ is the probability that the ith constraint is less than or equal to 0; r is the reliability aimed for by the ith constraint; and $\mu(h_j)$ is the mean value of the jth equation constraint.

The fundamental challenges of RBDO manifest in two dimensions, namely the issues of solution accuracy and efficiency. Wuchao [110] addressed these challenges by substituting the optimisation process with an interval analysis method grounded on Taylor expansion. This approach, tailored to the sensitivity of uncertainty in double nested inner layer optimisation, mitigates the need for double nested optimisation, substantially reducing computational burden during the optimisation process. Li [111] devised a method wherein the orientation vector is derived from the gradient of the response surface function. By rotating the original coordinate system based on the rotation matrix of the orientation vector and approximating the limit state function with new sample points in the new coordinate system, computational accuracy is enhanced, and the computational workload is reduced compared to traditional RSM, as evidenced by examples. Minrong [92] streamlined the determination of non-probabilistic reliability indexes in inner layer optimisation by framing it as seeking the minimum value of functional functions, capitalising on the geometric implications of non-probabilistic reliability indexes to enhance optimisation efficiency. Hailiang [112] combined the enhanced downhill simplex algorithm with response surface reliability analysis to refine the response surface approximation model. Through example analysis, it was demonstrated that this refined global optimisation algorithm exhibits high computational accuracy and commendable convergence capability.

1.4.3 Robust Optimisation Design Based on Reliability

RDO primarily addresses events proximate to the mean of the distribution, denoting small deviations around the norm. It represents an optimisation approach aimed at enhancing system stability and mitigating the impact of uncertainty on system performance. In contrast, RBDO focuses on extreme events occurring at the tails of the distribution. Here, the emphasis lies on assessing the influence of uncertainty in design variables and parameters on constraints, particularly in terms of the probability of constraint failure. By integrating these two approaches to uncertainty, the pursuit of system enhancement in both robustness and reliability is achievable, known as reliability-based robust design optimisation (RBRDO). This methodology simultaneously considers the impact of uncertainty on both optimisation objectives and constraints. Its mathematical expression is:

$$\begin{aligned} &\min \mu(F) \quad \text{or/and} \quad \sigma(F) \\ \text{s.t.} \quad &P(g_i \leqslant 0) \geqslant r, \quad i = 1, 2, \cdots, n \\ &\mu(h_j) = 0, \quad j = 1, 2, \cdots, m \end{aligned}$$

RBRDO represents an inevitable progression in the advancement of high-quality and high-precision design. Wang [113] conducted RBDO, RDO, and RBRDO separately, aiming to enhance system quality through reliability robust optimisation based on the kinematic design of traditional mechanisms, with cost and system motion accuracy (standard deviation and mean value of angular error) as the objectives. Na [114] combined reliability sensitivity theory with robust design methods to explore the reliability robust design challenges in gear transmission. Yichuan [115] established the parameter range of the machine tool cutting system as constraints, optimising the system to maximise stability and reliability while minimising the sum of squares of sensitivity for each system parameter as the objective function. This approach increases the reliability of the optimised system while reducing sensitivity to system parameters, thereby enhancing overall reliability and robustness. Liao [116] utilised the permissible reliability of the flexible hinge as a constraint, with hinge quality and reliability sensitivity as the objective function for RBRDO. This process decreases susceptibility to fluctuations in design variables while ensuring reliability requirements are met.

1.5 Status of Research on Optimal Ship Design Under Uncertainty

Ship optimisation design constitutes the core aspect of overall ship design, with related research typically founded on a simplified external environment. Crucial parameters, such as the ship's service speed, are often presumed to be definite quantities during the design process. However, in the actual operation of ships, uncertainty surrounding these parameters is inevitable. Neglecting this uncertainty and treating parameters as fixed values can result in design errors. Within the ship design process, where continuous iteration and mutual coupling occur, such errors tend to be gradually magnified, potentially compromising the robustness and reliability of final outcomes. Consequently, disregarding the impact of uncertainty may pose latent risks during the ship's operational phase. Therefore, designers must comprehensively consider various types of uncertainties at the design's outset to ensure that subsequent optimisation designs align with changes in relevant parameters.

Currently, numerous scholars have undertaken research concerning ship uncertainty-based design optimisation. Vasconcellos [117] focused on a high-speed catamaran in the Amazon River Basin, considering its operation under night conditions or during traffic congestion. Genetic algorithms were employed to optimise the catamaran, with speed reduction and sailing distance treated as uncertain variables. This approach aided designers in selecting variables during the initial design stages. Hou [118] investigated the underwater vehicle DATCOM, with drag force as the optimisation objective and DATCOM's scale parameters as uncertain quantities. The performance of DATCOM was evaluated using the 6-sigma design criterion to ensure the reliability of the underwater submersible amidst environmental disturbances and scale errors. Building on this, Hou utilised the energy efficiency design index (EEDI) as the optimisation objective, considering reliability design requirements under the influence of uncertainties. The 6-sigma criterion was applied to optimise the profile design of a large cargo-carrying bulk carrier. Compared to traditional deterministic optimisation, reliability optimisation design results exhibited better adaptability under environmental disturbances [119]. Papanikolaou explored the impact of uncertainty across multiple disciplines (drag, wave resistance, propulsion, economics, etc.) throughout the operational life cycle of a tanker, both in calm water and representative weather conditions. This study aimed to mitigate operational risks and assist potential investors in decision-making processes. Claus proposed a methodology aimed at progressively reducing the design space for uncertainty, thereby enhancing the designer's flexibility. This methodology utilised the complexity and loss of the design space as indicators of design performance. Claus et al. introduced a method for gradually reducing uncertainty in the design space and enhancing designer flexibility. This approach, employing complexity and loss of the design space as indices, proposed a double-layer nested optimisation method, facilitating designers in obtaining more information about space reduction during the initial design phases.

Moreover, Diez [120] has held an internationally prominent position in ship uncertainty-based design optimisation research. Diez's initial focus was on the RDO of ships. Initially, considering the stochastic nature of probabilistic parameters, Diez investigated the entire lifecycle cost of the vehicle using a multidisciplinary design optimisation approach, treating economically relevant parameters as uncertainties to enable robust optimisation of the vehicle [121]. Subsequently, Diez utilised a particle swarm optimisation (PSO) algorithm to robustly optimise the design solution of a bulk carrier by adopting the mean and variance of the objective function as the optimisation objectives [121]. Building upon this, a two-stage ship optimal design method, accounting for uncertainty, was proposed, with the mathematical expectation of the total cost function serving as the optimisation objective [122]. Following this, Diez [123] systematically summarised the application of this method in both fields, aiding designers in early-stage decision-making through the RDO concept. Diez et al. [124] subsequently extended the method's application to the multidisciplinary RDO of the keel fin of a racing yacht, demonstrating that the RDO scheme exhibited less sensitivity to external disturbances compared to traditional multidisciplinary design optimisation results. Additionally, Diez [125] introduced a robust optimisation method based on an agent model to enhance the efficiency of multi-disciplinary RDO while ensuring parameter accuracy, applying it to the hydrostatic drag optimisation of the DTMB5415 standard model.

Additionally, Diez [126] also implemented the reliability-based optimisation design method in the ship domain. Initially, the PSO algorithm was utilised for RBRDO of a high-speed catamaran to assess ship resistance and manoeuvrability under random sea conditions. Subsequently, considering the hydrodynamic abilities of the ship amidst uncertainty [127], various optimisation strategies were applied to the RBRDO of a high-speed catamaran in the North Pacific Ocean. These strategies included multi-objective aleatory uncertainty optimisation under random waves and speed, multi-objective aleatory uncertainty optimisation under fixed sea state and speed, deterministic optimisation under fixed speed of single regular wave, single-objective stochastic optimisation under fixed speed of calm water, and single-objective deterministic optimisation under fixed speed of single regular wave. These approaches were facilitated through the adaptive agent model and spatial scaling down based on the Karhunen–Loeve (KL) expansion method [128]. Upon comparison of the optimisation results, it was observed that the first optimisation method proved the most effective, while the third method was deemed the most economical. Furthermore, Diez [129] optimised the profile of a monohull with the objective of mitigating the impact of ship waves on the shoreline.

In addition to the aforementioned studies, both local and international research on ship uncertainty modelling remains in its nascent stages. Papanikolaou [130], for instance, examined the effects of ship response and load uncertainty induced by waves on the ship's safe and efficient operation. Employing the cross-spectrum combination method, various hydrodynamic ship responses were modelled, providing designers with an efficient system to manage extensive data and random loads. This approach facilitates the validation of decisions during the design process by aiding in reasonable assumptions. Furthermore, Hou [131] explored four forms of uncertainty and their applications in interval robust optimisation, focusing on the Wigley standard ship model. Utilising the Energy Efficiency Operational Indicator (EEOI) as the optimisation objective, the study aimed to demonstrate the feasibility and superiority of this methodology. Brefort [132], on the other hand, investigated the impact of epistemic uncertainty in ship design, employing fuzzy theory to model uncertainty in the ship's initial design phase. By embedding expert knowledge in multidisciplinary optimisation, Brefort optimised the scale of surface taxiing boats with drag and seakeeping as the optimisation goals, achieving a balanced conflict between the two disciplines while minimising the influence of uncertainties. Diez [133] quantified cognitive uncertainty using Bayesian methods and modelled stochastic uncertainty through PDFs. This approach was applied to optimise the profile of a container ship, with wave resistance as the primary optimisation objective, while simultaneously addressing robustness and reliability. Peri [134] gathered extensive operational data from a two-month data collection of five sister ships, deducing probability distributions of uncertainty quantities such as displacement and speed. After slight adjustments to these distributions, uncertainty optimisation was conducted to enhance ship type optimisation and reduce operational costs. Similarly, Claus [135] observed a fleet of 15 cargo ships over 18 months to model uncertainty variables based on this data. Utilising a Computer-Aided Engineering (CAE) platform integrated with Computational Fluid Dynamics (CFD) software, Claus evaluated ship programmes throughout their lifecycle. The results illustrated that considering uncertainties throughout the ship's lifecycle, supply chain, and actual operations can notably reduce freight rates and the EEOI.

There are relatively fewer studies on uncertainty analysis. Klasen [136] and Good [137], successively utilised the Confidence of Success (COS) method based on the mean value approach for transferring uncertainty analysis. They applied this method to optimise the conceptual design of naval equipment. However, these studies primarily considered uncertainty to guide the selection of the Pareto front by designers, indicating that the handling of uncertainty remains at a relatively elementary stage. Vlahopoulos [138–141] introduced uncertainty into ship multidisciplinary design, employing parallel networks to simultaneously optimise robustness and reliability. The optimisation objectives were the mean and standard deviation of the response, with constraints transformed into probabilistic constraints under the required reliability. This approach was applied to optimise the design of bulk carriers. Similarly, Diez [142] addressed uncertainty analysis, initially proposing a method for transferring uncertainty analysis in integral form. They compared various forms of robust optimisation and validated the superiority of their approach through an example of robust optimisation for a bulk carrier. Subsequently, Diez focused on the resistance, heave, and trim of a high-speed catamaran in calm water [143, 144]. They reduced the design space using the KL expansion method, considering uncertainties in geometric scale and Fr. Uncertainty analysis was conducted using the MC method and Gaussian interpolation method to predict the mean, variance, PDF, and Cumulative Density Function (CDF) of the response. Results indicated that the Gaussian interpolation method efficiently accomplished the task, while the MC method, combined with an approximation model, provided confidence intervals and distributions of the response at a low computational cost. Furthermore, Stern et al. [145] discussed uncertainty analysis transfer issues for ship hydrodynamics, considering the lift and drag of the NACA0012 wing under fixed angles of attack with random Reynolds numbers. They also addressed the uncertainty analysis transfer problem of a catamaran's drag in calm water or waves, comparing the advantages and disadvantages of three uncertainty analysis transfer methods, namely the MC method, MC combined with approximate modelling, and the integral method. Diez [146] experimentally verified the accuracy and reliability of the aforementioned methods. Finally, considering the hydroelastic mechanics of the keel fin of a rowing boat, Diez [147] proposed a multidisciplinary robust optimisation framework with varying degrees of coupling between multidisciplinary analyses, combined with the MC method for uncertainty analysis transfer. They also employed design of experiments, the spline agent model, and the improved PSO algorithm to address computational efficiency issues within the framework.

1.6 Problems with Ship Optimisation Design Under Uncertainty

Research on uncertainty optimal design primarily concentrates on uncertainty modelling and uncertainty analysis transfer. In aerospace, aviation, and automotive fields, this research has reached a relatively advanced stage. The modelling of uncertain quantities has evolved from traditional probabilistic methods to mixed probabilistic and non-probabilistic approaches. Concurrently, due to designers' pressing need for robustness and reliability, uncertainty analysis transfer methods have rapidly developed and found extensive application in the industry.

In recent years, the concept of uncertainty optimisation design in the maritime sector has begun to emerge in the design process. However, the current state of ship uncertainty optimisation design remains in its early stages, with limited attention to the inherent challenges of uncertainty optimisation itself. Specifically, the following issues are prevalent in ship uncertainty optimisation design:

(1) Lack of classification method research on uncertainty factors: Presently, designers often lack a thorough analysis of the characteristics of uncertainty factors. Uncertainty factors across the entire lifecycle of a ship are typically treated as strong statistical variables without precise classification. This oversight results in a lack of a scientific foundation for selecting subsequent modelling and analysis transfer methods.
(2) Costly aleatory uncertainty analysis and transfer methods: Many ship designers still rely on the time-consuming MC method for analysis and transfer under aleatory uncertainty. However, ship uncertainty optimisation design frequently demands numerous uncertainty analyses and simulation calculations, significantly increasing computational costs and complicating the use of the MC method.
(3) Inaccuracy and inadequacy of epistemic uncertainty modelling methods: Due to the absence of research into categorising uncertainty quantities, designers often struggle to choose appropriate modelling methods. Currently, most studies on epistemic uncertainty continue to employ traditional probabilistic approaches to modelling. However, due to insufficient information, variables and their corresponding parameters are often assumed with varying degrees of approximation. Obtaining precise probability density distributions from reality is challenging, rendering probabilistic methods less accurate and reasonable for modelling cognitive uncertainty.
(4) Absence of analytical transfer methods under mixed aleatory and epistemic uncertainty: While aleatory uncertainty analysis transfer methods based on probabilistic approaches have a long history and comprehensive mathematical theory, the foundation for epistemic uncertainty analysis transfer based on non-probabilistic methods is weaker. Moreover, there is a lack of methods that comprehensively consider both types of uncertainty. Developing unified frameworks for analysing both aleatory and epistemic uncertainty is essential for advancing research in this area.

In conclusion, aleatory uncertainty theory proves insufficient in effectively addressing uncertainty optimisation challenges, particularly in scenarios involving epistemic uncertainty or the coexistence of both forms. Consequently, there is a pressing need for novel theories and methodologies capable of tackling such optimisation dilemmas. Hence, it becomes imperative to account for both aleatory and epistemic uncertainty, thereby establishing a mathematical model for ship optimisation design under multi-source uncertainty. This entails conducting systematic theoretical investigations and numerical analyses on uncertainty modelling and analysis transfer. By doing so, a dependable foundation and efficient computational techniques can be provided for ship optimisation design amidst aleatory and epistemic uncertainty. Moreover, this endeavour aims to enrich and refine the existing theoretical framework for ship optimisation design in uncertain conditions. Ultimately, this study strives to furnish a robust basis and effective computational strategies for ship optimisation design under the influence of aleatory and epistemic uncertainty.

References

1. Feng W, Liu BG, Ding H, et al. Review of uncertain nonparametric dynamic modeling. J Vibrat Shock. 2020;39(5):1–9 (in Chinese).
2. Wang Y. Uncertainty-based optimization method andits application in aircraft design. Nanjing: Nanjing University of Aeronautics and Astronautics; 2010. (in Chinese).
3. Fang L. Interval non-probabilistic reliability analysis of uncertainty structure based on affine algorithm. Xi'an: Xidian University; 2014. (in Chinese).
4. De Laurentis DA, Mavris DN. Uncertainty modeling and management in multidisciplinary analysis and synthesis. In: The 38th aerospace sciences meeting and exhibit. Reno;2000. p. 422.
5. Walton MA. Managing uncertainty in space systems conceptual design using portflo theory. Boston: Massachusetts Instiute of Technology; 2002.
6. Batill SM, Renaud JE, Gu X. Modeling and simulation uncertint in multisciliry dsin optimization. In: The 8th AIAA/USAF/NASA/ISSMO symposium on multidisciplinary analysis and optimization. Long Beach;2000. p. 4803.
7. Du X, Chen W. A methodology for managing the effect of uncertainty in simulation-based design. AIAA J. 2000;38(8):1471–8.
8. De Weck O, Claudia E, Clarkson J. A cassification of uncertinty for erly prduet and stem design. In: International conference on engineering design. Paris;2007. p. 480.
9. Uebelhart SA, Millery DW, Blaurock C. Uncertainty characterization in integrated modeling. In: The 46th AIAA/ASME/ASCE/AHS/ASC structures, structural dynamics & materials. Austin;2005. p. 2142.
10. McKay MD, Morrison JD, Upton SC. Evaluating predietion uncertainty in simulation models. Comp Phys Commun. 1999; 17(1–2):44–51.
11. Oberkampf WL, Deland SM, Rutherford BM. Estimation for total uncertainty in modeling and simulation. Sandia: Sandia National Laboratories;2000.
12. Oberkampf WL, Helton JC, Sentz K. Mathematical representation of uncertainty. Seatle: Non-Deterministic Approaches Forum; 2001. p. 1645.
13. Klir GJ. Uncertainty and information measures for imprecise probabilities: an overview. In: The Ist International symposium on imprecise probabilities and their applications. Ghent;1999. p. 2691.
14. Walley P. Towards a unified theory of imprcise probability. Int J Prxorim Reason. 2000;24(2–3):125–48.

15. Haimes YY, Barry T, Lambert JH. When and how can you specify a probability distribution when you don't know much. Risk Anal. 1994;14(5):661–706.
16. Hattisl D, Burmate DE. Assessment of variability and uncertainty distributions for pracical risk analyses. Risk Anal. 1994;14(5):713–30.
17. Rice JA. Mathematical stistics and data analyis. 3rd ed. California: Duxbury Press; 2006.
18. Marhadi K, Venkataraman S, Pai SS. Quantifying uncertainty in statistical distribution of smallı sample data using Bayesian inference of unbounded Johnson distribution. In: The 49th AIAA/ASME/ASCE/AHS/ASC structures, structural dynamics, and materials conference. Schaumburg;2008. p. 1810.
19. Yao DL. System reliability method and research of small sample. Shenyang: Shenyang University of Aeronautics and Astronautics; 2011. (in Chinese).
20. Jiang C. Theories and algorithms of uncertain optimization based on interval. Changsha: Hunan University; 2008. (in Chinese).
21. Qiao XZ. On reliability analysis and optimizationof uncertain structures. Xi'an: Xidian University; 2009. (in Chinese).
22. Jin HL. Dynamics analysis for uncertain structures. Xi'an: Xidian University; 2014. (in Chinese).
23. Ben-Haim Y, Elishakoff I. Convex models of uncertainty in applied mechanics. Amsterdam: Elsevier; 1990.
24. Bi RG. Non-probabilistic uncertainty convex model and reliabilityanalysis method incorporating correlation. Changsha: Hunan University; 2015. (in Chinese).
25. Wang P, Zhang CP. Method of identifying dynamic uncertain parameter intervals withimproved parallelepiped convex model. J Vibrat Eng. 2019;32(1):97–106 (in Chinese).
26. Xie L. Research on structural uncertainty analysis method based on polygonal convex set model. Changsha: Hunan University; 2019. (in Chinese).
27. Zadeh LA. Fuzzy sets as a basis for a theory of possibility. Fuzzy Sets Syst. 1978;1(1):3–28.
28. Zadeh LA. Fuzzy sets. Inf Control. 1965;8(3):338–53.
29. Maglaras G, Nikolaidis E, Haftka RT, et al. Analytical-experimental comparison of probabilistic methods and fuzzy set based methods for designing under uncertainty. Struct Multidiscip Optim. 1997;13(2–3):69–80.
30. Nikolaidis E, Chen S, Cudney H, et al. Comparison of probability and possibility for esign against catastrophic failure under uncertainty. J Mech Des. 2004;126(3):386–94.
31. Braibant V, Oudshoorn A, Boyer C, et al. Non-deterministic possibilistic approaches for structural analysis and optimal design. AIAA J. 1999;37(10):1298–303.
32. He L, Huang H, Du L, et al. A review of possibilistic approaches to reliability analysis and optimization in engineering design. In: Jacko J. Human-Computer Interaction, Part IV. New York: Springer-Verlag;2007. p. 1075–84.
33. Langley RS. Unified approach to probabilistic and possibilistic analysis of uncertain systems. J Eng Mech. 2000;126(11):1163–72.
34. Shin Y, Wu Y. A hybrid possibilistic-probabilistic analysis framework for uncertainty management. In: The 44th AIAA/ASME/ASCE/AHS structures, structural dynamies, and materials. Norfolk;2003. p. 1573.
35. Du L, Choi KK, Youn BD, et al. Possibility-based design optimization method for design problems with both statistical and fuzzy input data. In: The 6th world congresses of structural and multidisciplinary optimization. Rio de Janeiro;2005. p. 216–37.
36. Wiener N. The homogeneous chaos. Am J Math. 1938;60(4):897–936.
37. Eldred MS, Burkardt J. Comparison of non-intrusive polynomial chaos and stochastic collocation methods for uncertainty quantification. In: The 47th AIAA aerospace sciences meeting including the new horizons forum and aerospace exposition. Orlando;2009. p. 128–9.
38. Eldred MS. Recent advances in non-intrusive polynomial chaos and stochastic collocation methods for uncertainty analysis and design. In: The 50th AIAA/ASME/ASCE/AHS/ASC structures, structural dynamics, and materials conference. Palm Springs;2009. p. 2274.
39. Helton JC, Johnson JD, Sallaberry CJ, et al. Survey of sampling-based methods for uncertainty and sensitivity analysis. Sandia: Sandia National Laboratories;2006.

40. Landau DP, Binder K. A guide to Monte Carlo simulations in statistical physics. 2nd ed. NewYork: Cambrige University Press; 2005.
41. Christian R, George C. Monte Carlo statistical methods. 2nd ed. London: Springer; 2004.
42. Ang GL, Tang WH. Optimal importance-sampling density estimator. J Eng Mech. 1992;118(6):1146–63.
43. Hinrichs A. Optimal importance sampling for the approximation of integrals. J Complex. 2010;26(2):125–34.
44. Cao Y, Hussaini MY, Zang TA. On the exploitation of sensitivity derivatives for improving sampling methods. In: The 44th AIAA structures, structural dynamics and mechanics conference. Norfolk;2003. p. 1656.
45. Augustin F, Gilg A, Paffrath M, et al. Polynomial chaos for the approximation of uncertainties: chances and limits. Eur J Appl Math. 2008;19:149–90.
46. Oladyshkin S, Class H, Helmig R, et al. An integrative approach to robust design and probabilistic risk assessment for CO2 storage in geological formations. Comput Geosci. 2011;15(3):565–77.
47. Buzzard GT. Global sensitivity analysis using sparse grid interpolation and polynomial chaos. Reliab Eng Syst Saf. 2012;107:82–9.
48. Sandoval EH, Anstett-Collin F, Basset M. Sensitivity study of dynamie systems using polynomial chaos. Reliab Eng Syst Saf. 2012;104:15–26.
49. Sudret B. Global sensitivity analysis using polynomial chaos expansions. Reliab Eng Syst Saf. 2008;93(7):964–79.
50. Oladyshkin S, De Barros FPJ, Nowak W. Global sensitivity analysis: a flexible and efficient framework with an example from stochastic hydrogeology. Adv Water Resour. 2012;37:10–22.
51. Ghanem RG, Spanos PD. Stochastic finite elements: a spectral approach. New York: Courier; 2003.
52. Ghanem RG, Spanos PD. A stochastic Galerkin expansion for nonlinear random vibration analysis. Probab Eng Mech. 1993;8(3–4):255–64.
53. Xiu D, Karniadakis GE. Modeling uncertainty in flow simulations via generalized polynomial chaos. J Comp Phys. 2003;187(1):137–67.
54. Keese A, Matthies HG. Sparse quadrature as an alternative to Monte Carlo for stochastic finite element techniques. In: Proceedings in applied mathematics and mechanics. Berlin;2003. p. 493–4.
55. Isukapalli SS, Roy A, Georgopoulos PG. Stochastic response surface methods (SRSMs) for uncertainty propagation: application to environmental and biological systems. Risk Anal. 1998; 18(3):351–63.
56. Li H, Zhang D. Probabilistic collocation method for flow in porous media: comparisons with other stochastic methods. Water Resourc Res. 2007; 43(9):97–109.
57. Wackernagel H. Multivariate geostatistics: an introduction with applications. Berlin: Springer; 2003.
58. Ditlevsen O, Madsen HO. Structural reliability methods. New York: Wiley; 1996.
59. Xiu D, Karniadakis GE. The Wiener-Askey polynomial chaos for stochastic differential equations. SIAM J Sci Comput. 2002;24(2):619–44.
60. Knio OM, Najm HN, Ghanem RG. A stochastic projection method for fluid flow: I. basic formulation. J Comp Phys. 2001; 173(2): 481–511.
61. Le Maitre OP, Reagan MT, Najm HN, et al. A stochastic projection method for fluid flow: II. Random process. J Comput Phys. 2002;181(1):9–44.
62. Askey R, Wilson J. Some basic hypergeometric polynomials that generalize Jacobi polynomials Memoirs Amer. Am Math Soc. 1985; 319.
63. Wan X, Karniadakis GE. Multi-clement generalized polynomial chaos for arbitrary probability measures. SIAM J Sci Comput. 2006;28(3):901–28.
64. Wan X, Karniadakis GE. Error control in multi-element generalized polynomial chaos method for elliptic problems with random coefficients. Commun Comp Phys. 2009;5(2–4):793–820.

65. Breitung K. Asymptotic approximations for probability integrals. Probab Eng Mech. 1989;4(4):187–90.
66. Song BF. A numerical integration method for computing structural system reliability. Comput Struct. 1990;36(1):65–70.
67. Sakamoto J, Mori Y, Sekioka T. Probability analysis method using fast Fourier transform and its application. Struct Saf. 1997;19(1):21–36.
68. Penmetsa RC, Grandhi RV. Adaptation of fast Fourier transformations to estimate structural failure probability. Finite Elem Anal Des. 2003;39(5–6):473–85.
69. Chen X, Lind NC. Fast probability integration by three-parameter normal tail approximation. Struct Saf. 1982;1(4):269–76.
70. Kim NH, Ramu P. Tail modeling in reliability-based design optimization for highly safe structural systems. In: The 47th AIAA/ASME/ASCE/AHS/ASC structures, structural dynamics, and materials conference. Newport;2006. p. 1825.
71. Acar E, Rais-Rohani M, Eamon CD. Reliability estimation using dimension reduction and extended generalized lambda distribution. The 49th AIAA/ASME/ASCE/AHS/ASC structures, structural dynamics, and materials conference. Schaumburg;2008. p. 1893.
72. Rahman S, Xu H. A univariate dimension-reduction method for multi-dimensional integration in stochastic mechanics. Probab Eng Mech. 2004;19(4):393–408.
73. Youn BD, Zhimin X, Wells L, et al. Enhanced dimension-reduction (eDR) method for reliabilitybased robust design optimization. In: The 11th AIAA/ISSMO multidisciplinary analysis and optimization conference. Portsmouth;2006. p. 6977.
74. Rackwitz R. Reliability analysis: a review and some perspectives. Struct Saf. 2001;23:365–95.
75. Hohenbichler M, Gollwitzer S, Kruse W, et al. New light on first- and second- order reliability methods. Struct Saf. 1987;4(4):267–84.
76. Zhao Y, Ono T. A general procedure for first/second-order reliability method(FORM/SORM). Struct Saf. 1999;21(2):95–112.
77. Grandhi RV, Wang L. Higher-order failure probability calculation using nonlinear approximations. Comput Methods Appl Mech Eng. 1999;168(1–4):185–206.
78. Zhao Y, Ono T, Kato M. Second-order third-moment reliability method. J Struct Eng. 2002;128(8):1087–90.
79. Zhao Y, Ono T. Moment methods for structural reliability. Struct Saf. 2001;23(1):47–75.
80. Du X, Chen W. A most probable point based method for uncertainty analysis. In: Proceedings of DETC'00, ASME 2000 design engineering technical conferences and computers and information in engineering conference. Baltimore;2000. p. 14263.
81. Elishakoff I, Hasofer AM. Exact versus approximate determination of structural reliability. Appl Sci Res. 1987;44(3):303–12.
82. Breitung K. Asymptotic approximations for multinormal integrals. J Eng Mech. 1984;110(3):357–66.
83. Ditlevsen O. Generalized second moment reliability index. Struct Mach. 1979;7(4):435–51.
84. Engelund S, Rackwit R. A benchmark study on importance sampling techniques in structural reliability. Struct Saf. 1993;12(4):255–76.
85. Bucher CG. Adaptive sampling: an iterative fast Monte Carlo method. Struct Saf. 1988;5(2):119–26.
86. Au S, Beck JL. Estimation of small failure probabilities in high dimensions by subset simulation. Probab Eng Mech. 2001;16(4):263–77.
87. Koutsourelakis PS, Pradlwarter HJ, Schueller GI. Reliability of structures in high dimensions, part I: algorithms and applications. Probab Eng Mech. 2004;19(4):409–17.
88. Ditlevsen O, Melchers RE, Gluver H. General multi-dimensional probability integration by directional simulation. Comput Struct. 1990;36(2):355–68.
89. Rajashekhar MR, Ellingwood BR. A new look at the response surface approach for reliability analysis. Struct Saf. 1993;12(3):205–20.
90. Ben-Haim Y. A non-probabilistic concept of reliability. Struct Saf. 1994;14(4):227–45.
91. Guo SX, Lv ZZ, Feng YS. A non-probabilistic model of structural reliability based on interval analysis. Chinese J Comp Mech. 2011;18(1):56–60 (in Chinese).

92. Wang MR. A dissertation submitted in partial fulfillment of the requirements for the degree of doctor of philosophy in engineering. Xi'an: Xidian University; 2019. (in Chinese).
93. Jiang T. Research on non-probabilistic algorithms forreliability of structures and systems. Xi'an: Xidian University; 2006. (in Chinese).
94. Beale EML. On minizing a convex function subject to linear inequalities. J Roy Stat Soc: Ser B (Methodol). 1955;17(2):173–84.
95. Stougie L. Design and analysis of methods for stochastic integer programming. Amsterdam: University of Amsterdam; 1985.
96. Hené TS, Dua V, Pistikopoulos EN. A hybrid parametric/stochastic programming approach for mixedinteger nonlinear problems under uncertainty. Ind Eng Chem Res. 2002;41(1):67–77.
97. Bastin F. Nonlinear stochastic programming. Namur: University of Namur; 2001.
98. Bastin F. Trust-region algorithms for nonlinear stochastic programming and mixed logit models. Namur: University of Namur;2004.
99. Mulvey JM, Vanderbei RJ, Zenios SA. Robust optimization of large-scale systems. Oper Res. 1995;43(2):264–81.
100. Chen X, Sim M, Sun P. A robust optimization perspective on stochastic programming. Oper Res. 2007;55(6):1058–71.
101. Zhang Y, Monder D, Forbes JF. Real-time optimization under parametric uncertainty: a probability constrained approach. J Proc Control. 2002;12(3):373–89.
102. Kadam JV, Schlegel M, Srinivasan B, et al. Dynamic optimization in the presence of uncertainty from off-line nominal solution to measurement-based implementation. J Proc Control. 2007;17(5):389–98.
103. Ruszczynski A, Shapiro A. Handbooks in operations research and management science: stochastic programming. Amsterdam: Elsevier; 2003.
104. Kall P, Wallace SW. Stochastic programming. Chichester: John Wiley and Sons; 1994.
105. Delgado M, Verdegay JL, Vila MA. A general model for fuzzy linear programming. Fuzzy Sets Syst. 1989;29(1):21–9.
106. Zhang DS, Zhang X, Wang H, et al. Robust optimization of powertrain mounting system based on Taguchi method. Automotive Eng. 2014; 36(4):475–480+514. (in Chinese)
107. Wang T. Robust optimization design on axlebox suspension parameters of high-speed passenger train. Changsha: Central South University; 2014. (in Chinese).
108. Sun GY, Li GY, Yan XL, et al. Application research of multi-objective robust optimization for desigh and manufacture in vehicle body. Chinese J Solid Mech. 2011;32(2):186–96 (in Chinese).
109. Liu YL, Chen WL, Bao YD, et al. Non-probabilistic robust optimization for polygon blank shape based on interval analysis. Journal of Mechanical Engineering. 2014;50(10):65–71 (in Chinese).
110. Qi WC, Qiu ZP. Non-probabilistic reliability-based structural design optimization based on the interval analysis method. 2013; 43(1):85–93. (in Chinese)
111. Li SH. Analysis of structure reliability for hybrid uncertainty parameters. Xi'an: Xidian University; 2014. (in Chinese).
112. Su HL. Research on reliability method under uncertainty and its application in automotive structure design. Guangzhou: South China University of Technology; 2020. (in Chinese).
113. Wang XQ. System reliability analysis and robust designoptimization of elastic mechanism. Shenyang: Northeastern University; 2014. (in Chinese).
114. Zhou N. Theory research on reliability-based robust design of gear transmission with multiple failure modes. Shenyang: Northeastern University; 2012. (in Chinese).
115. Xu YC. Chatter stability analysis and reliability robust optimization design of machining system. Shenyang: Northeastern University; 2015. (in Chinese).
116. Liao TJ. Structural reliability sensitivity and reliability robust design for compliant mechanism. Jiangxi: Jiangxi University of Science and Technology; 2018. (in Chinese).
117. Vasconcellos JM, Moraes HB. Multi objective optimization under uncertainty for catamaran preliminary design. In: The 11th international conference on fast sea transportation. Hawaii;2011.p. 572–8.

118. Hou YH, Xiao L, Xu YM. AUV hull lines optimization with uncertainty parameters based on six sigma reliability design. Int J Naval Arch Ocean Eng. 2018;10(4):499–507.
119. Hou YH, Liang X, Mu XY. Hull lines reliability-based optimisation design for minimum eedi. Brodogradnja. 2018;69(2):17–33.
120. Diez M, Iemma U. Robust optimization of aircraft life-cycle costs including the cost of community noise. In: The 13th AIAA/CEAS aeroacoustics conference. Roma;2007. p. 3668.
121. Diez M, Peri D. Global optimization algorithms for robust optimization in naval design. In: International conference on computer applications and information technology in the maritime industries. Compit;2009. p. 128.
122. Diez M, Peri D. Two-stage stochastic programming formulation for ship design optimisationunder uncertainty. Ship Technol Res. 2010;57(3):172–81.
123. Diez M, Peri D, Campana EF, et al. Robust decision making in aerial and marine vehicles optimization: a designer's viewpoint. Enterprise Risk Manag. 2010;2(1):68–86.
124. Diez M, Peri D, Fasano G, et al. Hydroelastic optimization of a keel fin of a sailing boat: a multidisciplinary robust formulation for ship design. Struct Multidiscip Optim. 2012;46(4):613–25.
125. Leotardi C, Campana EF, Diez M. Resistance reduction of a military ship by variableaccuracy metamodel-based multidisciplinary robust design optimization. In: Vi International conference on computational methods for coupled problems in science and engineering. Venice;2015. p. 952–63.
126. Pellegrini R, Serani A, Diez M, et al. Multi-objective extensions of the deterministic particle swarm algorithm for RBRDO in ship design: a parametrie study. In: Numercial towing tank symposium. Marstrand; 2014. p. 28–30.
127. Pellegrini R, Campana EF, Diez M, et al. Application of derivative-free multi-objective algorithms to reliability-based robust design optimization of a high-speed catamaran in real ocean environment. Eng Optim IV. 2014;15:64.
128. Diez M, Campana EF, Stern F. Development and evaluation of hull-form stochastic optimization methods for resistance and operability. In: FAST 2015, international conference on FAST sea transportation. Washington DC;2015. p. 21.
129. Peri D, Diez M. Robust design optimization of a monohull for wave wash minimization. In: International conference on computational methods in marine engineering. Hamburg;2013. p. 89–100.
130. Papanikolaou A, Mohammed EA, Hirdaris SE. Stochastic uncertainty modelling for ship design loads and operational guidance. Ocean Eng. 2014;86(1):47–57.
131. Hou YH. Hull form uncertainty optimization design for minimum EEOI with influence of different speed perturbation types. Ocean Eng. 2017;140:66–72.
132. Brefort D, Singer DJ. Managing epistemic uncertainty in multi-disciplinary optimization of a planing craft. In: Marine Design XIII. New York;2018. p. 255–65.
133. Diez M, Peri D. Optimal hull-form design subject to epistemic uncertainty. Ship Technol Res. 2012;59(1):14–20.
134. Peri D. Robust design optimization for the refit of a cargo ship using real seagoing data. Ocean Eng. 2016;123:103–15.
135. Claus LR, Collette MD. An optimization framework for design space reduction in early stage design under uncertainty. New York: CRC Press;2018.
136. Klasen E. Confidence of success in multi-criteria optimization of multi-disciplinary ship design models. Blacksburg: Virginia Polytechnic Institute and State University; 2005.
137. Good NA. Multi-objective design optimization considering uncertainty in a multidisciplinary ship synthesis model. Blacksburg: Virginia Polytechnic Institute and State University; 2006.
138. Vlahopoulos N, He J. Application of a multidiscipline design optimization approach fordesigning the thermal protection system of an Apollo type vehicle under uncertainty. In: JANNAF Meeting. Orlando;2008. p. 1341–55.
139. Shari H, Nickolas V. Introducing uncertainty in multidiscipline ship design. Nav Eng J. 2010;122(2):41–52.

140. Hannapel S, Vlahopoulos N. Robust and reliable multidiscipline ship design. In: AIAA/ ISSMO multidisciplinary analysis optimization conference. Fort Worth;2013. p. 341–57.
141. Hannapel SE. Development of multidisciplinary design optimization algorithms for ship design under uncertainty. Ann Arbor: University of Michigan; 2012.
142. Diez M, Peri D. Robust optimization for ship conceptual design. Ocean Eng. 2010;37(11):966–77.
143. Kandasamy M, Peri D, Tahara Y, et al. Simulation based design optimization of waterjet propelled Delft catamaran. Int Shipbuild Prog. 2013;60:219–27.
144. Diez M, He W, Campana EF, et al. Uncertainty quantification of Delft catamaran resistance, sinkage and trim for variable Froude number and geometry using metamodels, quadrature and Karhunen-Loève expansion. J Mar Sci Technol. 2014;19(2):143–69.
145. Stern F, Volpi S, Gaul NJ, et al. Development and assessment of uncertainty quantification methods for ship hydrodynamics. In: AIAA aerospace sciences meeting. 2017. p. 1–9.
146. Diez M, Broglia R, Durante D, et al. Validation of uncertainty quantification methods for highfidelity CFD of ship response in irregular waves. In: AIAA acrospace sciences meeting. Texas;2017. p. 367–82.
147. Leotradi C, Diez M, Serani A, et al. A framework for efficient simulation-based multidisciplinary robust design optimization with application to a keel fin of a racing sailboat. In: International conference on engineering and applied sciences optimization. Athens;2014. p. 734–57.

Chapter 2
Classification of the Uncertainty

The proper classification of uncertain factors serves as the cornerstone for uncertainty modelling. Present research commonly categorises uncertain factors as robust statistical variables, resulting in discrepancies in subsequent modelling approaches and computational outcomes. This chapter adopts the statistical inference method, combining parameter estimation with the goodness-of-fit method. It proposes an uncertainty classification technique to precisely identify the variable type to which uncertain factors belong, establishing the groundwork for subsequent uncertainty analysis and propagation.

2.1 Types of Uncertainty

Typically, aleatory uncertainty and epistemic uncertainty are categorised into three types of uncertain variables: robust statistical variables denoted as X, sparse variables represented by Y, and interval variables labelled as Z.

1. Statistical variables X

If the random distribution type of a variable X can be easily obtained or an accurate statistical model of the variable can be derived from a large dataset, then X is classified as a robust statistical variable, denoted as:

$$X \sim \vartheta(\boldsymbol{\xi}) \tag{2.1}$$

where ϑ represents a specific probability distribution such as normal or uniform distribution, and ξ represents the distribution parameters corresponding to the probability distribution. For instance, if ϑ represents a normal distribution, then $\xi = [\xi_1, \xi_2]$, where ξ_1 and ξ_2 represent the mean and standard deviation, respectively.

2. Sparse variables Y

Z. Liu et al., *Uncertainty-Based Ship Design Optimization*,
https://doi.org/10.1007/978-981-97-2973-9_2

If only a limited amount of point data or interval data is available for the uncertain variable, it becomes challenging to accurately fit the data with the PDF of a single distribution. Consequently, sparse variables Y can be characterised by utilising the weighted sum of PDFs of multiple distributions, as follows:

$$Y \sim w_i \vartheta_i(\xi^i) \tag{2.2}$$

where w_i, ϑ_i, ξ^i respectively denote the weight, PDF, and parameters corresponding to the ith distribution.

3. Interval variables Z

When data is insufficient to construct any probability distribution, the variable becomes an interval variable. It can be succinctly represented as:

$$Z \in [\underline{z}, \overline{z}] \tag{2.3}$$

where $\underline{z}$ and $\overline{z}$ respectively denote the upper and lower bounds of the interval variable.

In practical scenarios, parameters data may originate from various sources. When addressing uncertainty data from diverse sources, it becomes necessary to classify the uncertainty based on the available information. Specifically, categorising uncertainty into the aforementioned three types is essential, followed by the application of corresponding mathematical tools for modelling and subsequent processing.

2.2 Statistical Inference Methods

Presently, no mature method exists for classifying uncertain factors. This section employs statistical inference methods to address this issue. Statistical inference comprises two components: parameter estimation and hypothesis testing. The former entails deducing the probability distribution of uncertain factors and their parameters based on assumptions, while the latter involves testing the aforementioned assumptions and inferences. Utilising the aforementioned method, uncertain factors are categorised into three types of uncertain variables in Sect. 2.1. This classification serves as the groundwork for subsequent uncertainty modelling and analysis.

2.2.1 Parameter Estimation

Estimation encompasses two scenarios: one where the distribution form is unknown, and the other where the distribution form is known but its corresponding parameters are unknown. The former necessitates a substantial number of samples to construct an empirical distribution function that approximates the real distribution function. However, acquiring such a large number of samples poses a challenge for designers.

Consequently, a common approach is to assume that the variable conforms to a specific distribution form and then estimate its corresponding distribution parameters based on the sample. Frequently employed estimation methods include MLE and Bayesian estimation.

1. **MLE**

Let us consider a random variable X with a PDF represented by $f(x|\xi)$, where ξ represents a parameter vector. For X, there exists an independent sample set represented by $x_i(i = 1, 2, \cdots, m)$. The likelihood function can be expressed as:

$$L(\xi) = f(x_0|\xi)f(x_1|\xi)\cdots f(x_m|\xi) \tag{2.4}$$

Due to the potential issue of underflow resulting from continuous multiplication of probabilities, it is common practice to use the log-likelihood:

$$LL(\xi) = \ln L(\xi) = \sum_{i=0}^{m} \ln f(x_i|\xi) \tag{2.5}$$

When $m \to \infty$, we have:

$$\frac{1}{m+1}\sum_{i=0}^{m} \ln f(x_i|\xi) \to E[\ln f(X|\xi)] \tag{2.6}$$

where E denotes the expectation of the distribution, expressed as:

$$E[\ln f(X|\xi)] = \int g(x) \ln f(x|\varepsilon)\,dx \tag{2.7}$$

and $g(x)$ represents the true distribution of X.

The Kullback information $I[g; f(|\boldsymbol{\xi})]$ quantifies the expectation of the logarithmic difference between the original distribution and the approximating distribution. It precisely measures the amount of information lost in the approximating distribution, as:

$$I[g; f(|\xi)] = E[\ln g(X)] - E[\ln f(X|\xi)] = \int g(x) \ln \frac{g(x)}{f(x|\varepsilon)}\,dx \tag{2.8}$$

The value is non-negative, and the smaller it is, the closer the approximate distribution is to the true distribution. In formula (2.8), it equals 0 only when $g(x)$ is consistent with $f(x|\xi)$. Therefore, maximising $E[\ln f(X|\xi)$ entails finding a $f(x|\xi)$ close to $g(x)$, which is the essence of MLE. Intuitively, MLE aims to find a value that maximises the likelihood of the data (sample) occurring among all possible values of ξ. At this point, the MLE of the parameter ξ is:

$$\xi^* = \arg\max LL(\xi) \tag{2.9}$$

In this book, given that the data source comprises a mixture of point data and interval data, the general steps of the MLE solution vary slightly from the traditional method. The steps are as follows:

(1) Construction of the likelihood function

Assuming that the random variable satisfies a certain distribution, but parameters ξ of the distribution are unknown and need to be determined through m data points $x_i(i = 1, 2, \cdots, m)$. $f_x(x|\xi)$ and $F_x(x|\xi)$ are respectively the conditional PDF and the CDF with respect to the parameter.

For the likelihood function $L(\xi)$, it is positively related to the probability, but for any discrete point x_i, its probability is 0. For point data, take the interval $x \in [x_i - \frac{\varepsilon}{2}, x_i + \frac{\varepsilon}{2}]$. When ε is infinitesimal, the likelihood function of point data is:

$$L(\xi) \propto P\left(x \in \left[x_i - \frac{\varepsilon}{2}, x_i + \frac{\varepsilon}{2}\right]|\xi\right) = \int_{x_i-\frac{\varepsilon}{2}}^{x_i+\frac{\varepsilon}{2}} f_x(x|\xi)\mathrm{d}x = \varepsilon f_x(x_i|\xi) \propto f_x(x_i|\xi) \quad (2.10)$$

Therefore, the likelihood function $L(\xi)$ of point data is positively related to the conditional PDF $f_x(x_i|\xi)$ of the point. For multiple point data, assuming they are independent of each other:

$$L(\xi) \propto \prod_{i=1}^{m} f_x(x_i|\xi) \quad (2.11)$$

Similarly, for interval data $[a, b]$, the likelihood function is:

$$L(\xi) \propto P(x \in [a, b]|\xi) = \int_a^b f_x(x|\xi)\mathrm{d}x = F_x(b|\xi) - F_x(a|\xi) \quad (2.12)$$

For multiple interval data $[a_j, b_j](j = 1, 2, \cdots, n)$, assuming that they are independent of each other:

$$L(\xi) \propto \prod_{j=1}^{n} \left[F_x(b_j|\xi) - F_x(a_j|\xi)\right] \quad (2.13)$$

In summary, if there are m point data $x_i(i = 1, 2, \cdots, m)$ and n interval data $[a_j, b_j]$ of x, and all data are independent of each other, then:

$$L(\xi) \propto \left[\prod_{i=1}^{m} f_x(x_i|\xi)\right]\left[\prod_{j=1}^{n} \left[F_x(b_j|\xi) - F_x(a_j|\xi)\right]\right] \quad (2.14)$$

(2) Take the logarithm of the likelihood function, and then derive the derivatives of the parameters to obtain a set of likelihood equations

For parametric vectors $\xi = [\xi_1, \xi_2, \cdots, \xi_n]$, there are layer operators:

$$\nabla_\xi = \left[\frac{\partial}{\partial \xi_1}, \frac{\partial}{\partial \xi_2}, \ldots, \frac{\partial}{\partial \xi_n}\right]^{\mathrm{T}} \tag{2.15}$$

If the likelihood function is continuously differentiable, then the likelihood equations are:

$$\nabla_\xi LL(\xi) = \nabla_\xi \ln P(D|\xi) = \nabla_\xi \sum_{x \in D} \ln P(x|\xi) = 0 \tag{2.16}$$

(3) Solve the likelihood equations to obtain the maximum likelihood estimator ξ^*.

This book selects six common probability distributions as candidate distributions for MLE (See Table 2.1). The derivation of MLE for these six candidate distributions is provided in Appendix A.

2. **Bayesian Estimation**

The Bayesian theory posits that parameters are unobserved random variables, which themselves may also follow distributions, expressed as:

Table. 2.1 Candidate distributions

Distributions	PDF	Parameters
ϑ_1:Normal	$f(x) = \frac{1}{\sqrt{2\pi\sigma^2}} \exp\left(-\frac{(x-\mu)^2}{2\sigma^2}\right)$	μ σ
ϑ_2:Uniform	$f(x) = \begin{cases} \frac{1}{b-a}, a \le x \le b \\ 0, \quad x < a \, or \, x > b \end{cases}$	a b
ϑ_3:Exponential	$f(x) = \begin{cases} \lambda e^{-\lambda x}, x > 0 \\ 0, \quad x \le 0 \end{cases}$	λ
ϑ_4:Weibull	$f(x) = \frac{\beta}{\alpha}\left(\frac{x}{\alpha}\right)^{\beta-1} \exp\left[-\left(\frac{x}{\alpha}\right)^{\beta}\right]$	α β
ϑ_5:Extreme value	$f(x) = \frac{1}{\sigma} \exp\left\{-\frac{1}{\sigma}(x-\mu) - \exp[-\frac{1}{\sigma}(x-\mu)]\right\}$	μ σ
ϑ_6:Lognormal	$f(x) = \begin{cases} \frac{1}{\sqrt{2\pi\sigma^2}x} e^{-\frac{(\ln x-\mu)^2}{2\sigma^2}}, x > 0 \\ 0, \quad x < 0 \end{cases}$	μ σ

$$P(\xi|D)=\frac{P(\xi)P(D|\xi)}{P(D)} \tag{2.17}$$

where $P(\xi)$ is the prior probability of the parameter; $P(D|\boldsymbol{\xi})$ is the conditional probability or the likelihood of the sample D with respect to ξ; and $P(D)$ is the normalised evidence factor.

The prior probability $P(\xi)$ of parameters represents the frequency estimate of each parameter sample in the sample space. Regarding the conditional probability $P(D|\boldsymbol{\xi})$ of parameters ξ, assuming $P(D|\boldsymbol{\xi})$ has a definite form and is uniquely determined by the parameter vector ξ, the training set $P(D|\boldsymbol{\xi})$ is employed to estimate the parameters ξ, which is the MLE. $P(D)$ represents the probability of the observed evidence, which can be obtained by expanding the total probability formula, as follows:

$$P(D)=P(D|\xi)P(\xi)=\sum_{i=0}^{k}P(x_i|\xi)P(\xi) \tag{2.18}$$

$$\rho(D)=\int\rho(D|\xi)\rho(\xi)\mathrm{d}\xi \tag{2.19}$$

Typically, when there is ample data available for uncertain factors, and the data conforms to the criteria for strong statistical variables, employing MLE to estimate parameters via the frequentist method is feasible. However, when the available data is limited and described by sparse or imprecise information, indicating sparse variables, traditional MLE for single probability distribution estimation becomes impractical.

Bayesian theory employs the Bayesian model averaging method to derive the PDF. This approach involves comparing multiple probability models by assigning weights to candidate probability distributions, with the weights and distribution parameters estimated from available data. In considering N candidate distribution types, this method expresses the probability distribution $f_y(y)$ of the input variable Y as a weighted sum of candidate PDFs. Let $f_y^k(y|\xi_k)$ represent the PDF of the kth candidate distribution. For each PDF, ξ_k represents the vector of distribution parameters, with weight w_k. Utilising the Bayesian model averaging method, the PDF of Y is:

$$f_y(y|\xi,w)=\sum_{i=1}^{N}w_kf_y^k(y|\xi_k),\ k=1,2,\cdots,N \tag{2.20}$$

where $\sum_{i=1}^{N}w_k=1$.

The likelihood estimation method is utilised to estimate both the weights and distribution parameters, with the likelihood function being expressed as:

$$L(\xi,w)\propto\left[\prod_{i=1}^{m}f_y(y=y_i|\xi,w)\right]\left[\prod_{j=1}^{n}\int_{a_j}^{b_j}f_y(y_i|\xi,w)dy\right] \tag{2.21}$$

Hence, both ξ and w can be estimated using MLE. Uncertain parameters within the PDF can be determined through Bayesian inference. The uncertainty surrounding distribution types and parameters can be quantified via the Bayesian model averaging method, facilitating estimates of belief (the mean of w) and uncertainty (the standard deviation of w) regarding a specific distribution and its associated trust. One limitation of this approach is that it implicitly accounts for non-existent interactions when establishing the joint PDF of the weights and distribution parameters of all candidate distributions. Additionally, this method entails multi-dimensional integration and is computationally intricate, particularly when dealing with multiple candidate distributions simultaneously. For instance, if there are five candidate distributions, each with two parameters, the joint PDF will encompass 14 variables (four weights and ten distribution parameters), necessitating a 14-dimensional integral for estimating the uncertainty of both the distribution type and its parameters.

2.2.2 *Hypothesis Testing*

The model itself introduces uncertainty, particularly in its probabilistic form. This uncertainty arises from assumptions regarding the distribution type that fits the data, which is based on past experience, expert opinions, experimental data, and other sources. Consequently, it becomes imperative to assess the feasibility of the chosen probabilistic model.

To assess the suitability of a hypothesised distribution for a dataset, hypothesis testing methods are employed to gauge the goodness of fit of a statistical model. The goodness-of-fit test serves as a means to measure the consistency between an empirical frequency distribution and an assumed probability distribution, thereby aiding in the selection of an appropriate probability distribution.

Conducting a goodness-of-fit test on a sample can ascertain whether the observed distribution aligns with a particular probability distribution, such as the normal distribution. It helps in determining whether the observed sample was drawn from a specific probability distribution. Typically, the fit test involves certain assumptions:

$$\text{Null hypothesis}\, H_0 : F = F_0 \,\text{Alternative hypothesis}\, H_1 : F \neq F_0 \tag{2.22}$$

where F is the actual distribution function and F_0 is the hypothesised distribution function.

The empirical distribution is juxtaposed with the hypothesised distribution, and if the empirical distribution deviates significantly from the hypothesised distribution, the null hypothesis is typically rejected. The determination of whether to accept or reject the null hypothesis hinges upon the specific test method employed.

Through the steps outlined above, the goodness-of-fit test aids in determining whether the observed sample conforms to a specific distribution. This approach encompasses both graphical and statistical methods. Statistical techniques include the Pearson chi-square test, Kolmogorov–Smirnov test, Cramér-von-Mises criterion, and Anderson–Darling test, among others. These methods evaluate the goodness of fit by comparing the assumed probability distribution with the empirical probability distribution. The subsequent discussion will delve into the Anderson–Darling test method.

The Pearson chi-square test imposes stringent requirements on the sample size and lacks precise interval division criteria. While the Kolmogorov–Smirnov test addresses the interval division issue encountered in the chi-square test, it still depends on a large number of samples. Consequently, its testing accuracy diminishes when the sample size is small, rendering it less than ideal in such situations.

The Anderson–Darling test offers greater precision in testing with small sample sizes and finds widespread use in such scenarios. Primarily, it is employed to ascertain whether a dataset conforms to a normal distribution and to evaluate whether a hypothesised distribution aligns with the sample data. This assessment involves computing the quadratic distance between the empirical distribution function and the hypothesised distribution function, followed by comparing it to a critical value (CV) to determine the acceptance or rejection of the hypothesis.

For the sample point $x_i (i = 1, 2, \cdots, n)$, the distance between the hypothetical distribution $F(x)$ and the empirical distribution $F_n(x)$ is:

$$n \int_{-\infty}^{+\infty} [F_n(x) - F(x)] w(x) \mathrm{d}F \tag{2.23}$$

where $w(x)$ is the weight function and n is the number of sample points. In the Anderson–Darling test, the weight function $w(x)$ is defined as:

$$w(x) = \{F(x)[1 - F(x)]\}^{-1} \tag{2.24}$$

This weighting emphasises observations in the tail distribution, enhancing the test's sensitivity to outliers and its ability to detect shifts in the tail distribution. The specific formula for this is:

$$A_n^2 = AD = -n - \sum_{i=1}^{n} \frac{2i-1}{n} \left[\ln(F(x_i)) + \ln(1 - F(x_{i+n-1})) \right] \tag{2.25}$$

At this stage, the sample data points are arranged in ascending order.

The decision to accept or reject the adherence of the data to the hypothesised distribution is determined by assessing the relationship between the observed significance level (OSL) and CV. The OSL of the distribution is calculated as follows:

$$AD^* = AD\left(1 + \frac{4}{n} - \frac{25}{n^2}\right) \tag{2.26}$$

$$OSL = [1 + \exp(-0.48 + 0.78 \ln(AD^*) + 4.58 AD^*)]^{-1} \tag{2.27}$$

2.3 Classification Method of the Uncertainty

Assuming that the variables conform to six common candidate distributions for strong statistical variables and one mixed distribution for sparse variables, respectively, the parameter estimation of these seven distributions is conducted using the MLE method based on the known variable data. This process yields the CDF $F(x)$ of the distributions. Subsequently, the Anderson–Darling test is employed, as per the $F(x)$, to compute the OSL for each distribution. The OSL value is then compared with the CV. This comparison allows for the accurate determination of the type of variable to which the uncertainty factor belongs, thereby completing the classification of uncertainty (Fig. 2.1).

For a sample of uncertain variables $x_i (i = 1, 2, \cdots, n)$, the classification process unfolds as follows:

Step 1 Initially, presume the variable to be a strong statistical variable and let it adhere to six candidate distributions. Parameters of these distributions are then estimated using MLE. Subsequently, consider the variable as a sparse variable, with the distribution type being the weighted sum of candidate distributions. This is expressed through a Bayesian averaging model.
Step 2: Solve the CDF of the six candidate distributions and one weighted distribution.
Step 3: Calculate the Anderson–Darling distance and subsequently derive the OSL. Identify the maximum OSL value among the seven distributions and compare it with the CV.

(1) If the maximum OSL exceeds the CV and the corresponding distribution is a single probability distribution, then the variable is classified as a strong statistical variable and adheres to the probability distribution.
(2) If the maximum OSL surpasses the CV, and the corresponding distribution represents the weighted sum of distributions, the variable is deemed a sparse variable following a mixed distribution.
(3) If the maximum OSL falls below the CV, the variable is classified as an interval variable.

$$\underline{Z} = Z_{\min} - \psi \cdot \sigma_Z \tag{2.28}$$

$$\overline{Z} = Z_{\max} + \psi \cdot \sigma_Z \tag{2.29}$$

where σ_Z, $Z_{\min}$, $Z_{\max}$ are the standard deviation, minimum, and maximum values of the sample points $x_i (i = 1, 2, \cdots, n)$, and ψ is the expansion parameters of the interval.

It is important to highlight that ψ is set to 0, the upper and lower bounds of the interval represent solely the extremes of the existing training samples without incorporating any redundancy or prediction regarding the unknown samples. Consequently, this scenario may lead to a situation where the test sample possesses a value

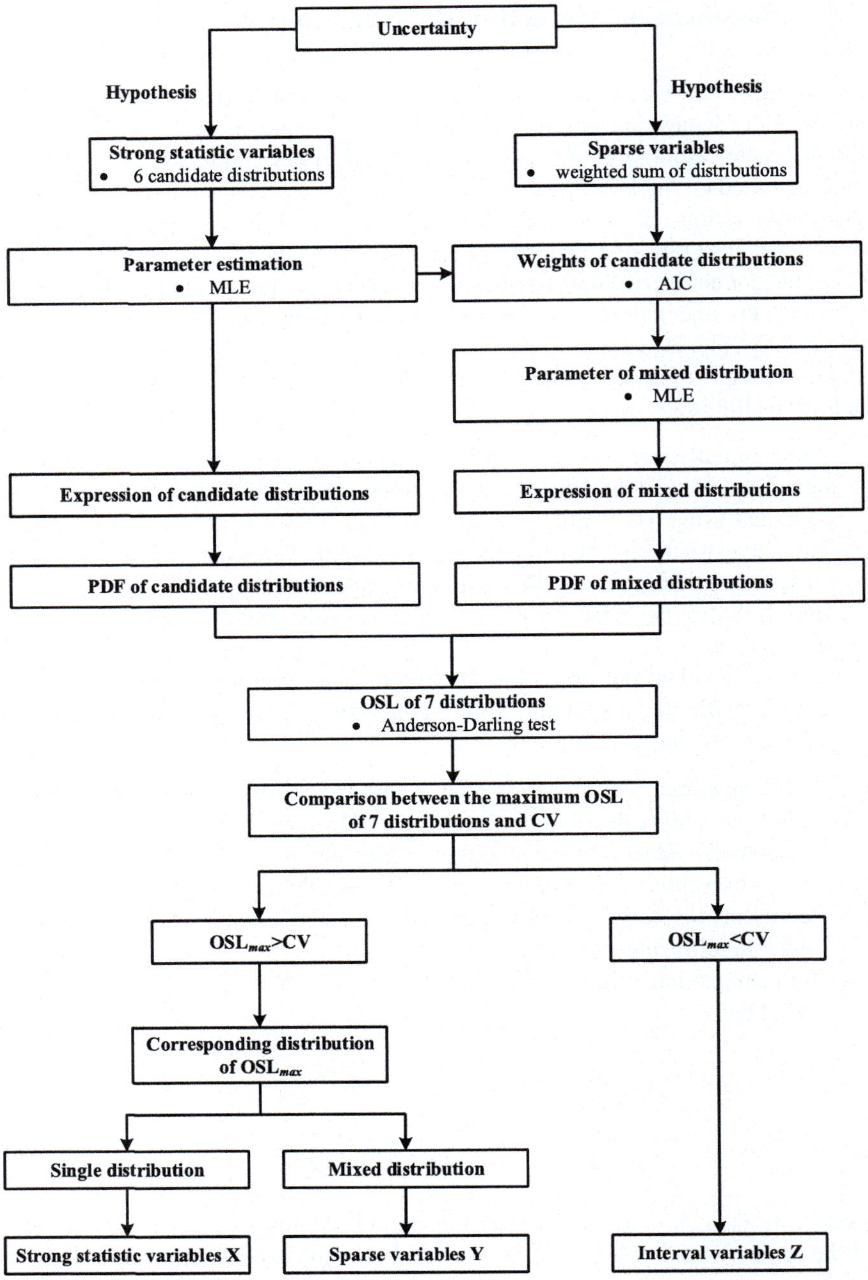

Fig. 2.1 Classification of the uncertainty

equal to either the maximum or minimum values of the training sample, with a probability of 0 assigned to that specific training sample. Such an outcome contradicts the actual circumstances.

2.4 Examples

An example involving a composite material has been chosen to validate the uncertainty classification method. The elastic parameters (E_1,E_2,V_{21},G_{12}) and the thickness t of each board have been identified as uncertain parameters. The experimental input data for these five parameters are presented in Table 2.2 [1, 2].

The data provided in Table 2.2 are insufficient for directly constructing the distribution function. Thus, the uncertainty classification method proposed in this chapter is employed to categorise the variable types of the aforementioned five parameters. Initially, assuming that these five variables are strong statistical variables and adhere to the six candidate distributions, their parameters are estimated, with the outcomes detailed in Table 2.3.

The variable is then considered to be a sparse variable represented as a weighted sum of candidate distributions, and its weights and parameters are determined. The PDF curves for the seven distributions are depicted in Fig. 2.2. Subsequently, the CDFs of the seven distributions for the five variables are computed. Through the utilisation of the Anderson–Darling test, the values of the OSLs are obtained for each variable, as presented in Table 2.4.

From Table 2.4, it is evident that for the five variables, the highest OSL of E_1 exceeds the CV (CV $= 0.2$ in this instance), and the probability distribution associated with this maximum OSL comprises a mixture of six probability distributions. Consequently, E_1 represents a sparse variable expressed through weighted probability distributions. Moreover, the highest OSL of E_2 surpasses the CV, indicating that the corresponding probability distribution is normal. Hence, E_2 is identified as a strong statistical variable following a normal distribution. Conversely, the highest

Table. 2.2 Experimental data of uncertain input parameters for composite materials

No	E_1/GPa	E_2/GPa	V_{21}	G_{12}/GPa	t/mm
1	126.91	8.67	0.30	4.85	4.76
2	128.28	9.07	0.32	4.98	4.98
3	130.39	9.08	0.33	5.00	5.13
4	130.63	9.18	0.33	5.13	5.14
5	132.00	9.21	0.33	5.16	5.27
6	132.19	9.28	0.33	5.25	5.28
7	135.30	9.39	0.34	5.34	5.29
8	137.33	9.73	0.35	5.45	5.33

Table. 2.3 Parameter estimation for uncertain variables

Variables	Parameters	Normal	Uniform	Exponential	Weibull	Extreme value	Lognormal
E_1	a	131.63	126.91	131.63	133.24	130.08	4.88
	b	3.21	137.33	0.01	41.71	2.75	0.02
	$Ln(L_{max})$	−20.68	−18.75	−47.04	−21.35	−20.58	−20.65
E_2	a	9.20	8.67	9.20	9.34	9.06	2.22
	b	0.28	9.73	0.11	33.44	0.28	0.03
	$Ln(L_{max})$	−1.22	−0.47	−25.75	−1.77	−1.87	−1.23
V_{21}	a	0.33	0.30	0.33	0.33	0.32	−1.11
	b	0.01	0.35	3.04	28.68	0.02	0.04
	$Ln(L_{max})$	23.01	23.97	0.90	23.28	21.72	22.89
G_{12}	a	5.15	4.85	5.15	5.24	5.05	1.64
	b	0.19	5.45	0.19	30.00	0.17	0.04
	$Ln(L_{max})$	2.09	4.09	−21.10	1.77	1.88	2.10
t	a	5.15	4.76	5.15	5.23	5.05	1.64
	b	0.18	5.33	0.19	41.41	0.21	0.04
	$Ln(L_{max})$	2.28	4.50	−21.11	3.47	0.80	2.13

OSL of V_{21} is less than the CV, indicating that V_{21} is an interval variable with an interval of $[0.295, 0.355]$. Furthermore, the highest OSL of G_{12} exceeds the CV, with the associated probability distribution being lognormal. Consequently, G_{12} is identified as a strong statistical variable following a lognormal distribution. Similarly, the highest OSL of t surpasses the CV, with the corresponding probability distribution being Weibull. Therefore, t is classified as a strong statistical variable adhering to a Weibull distribution.

Figure 2.3 illustrates that the chosen distribution function, following uncertainty classification, exhibits a superior fitting effect on the experimental data points. This effect is particularly notable in the high probability region, thereby corroborating the accuracy of the uncertainty classification method proposed in this chapter.

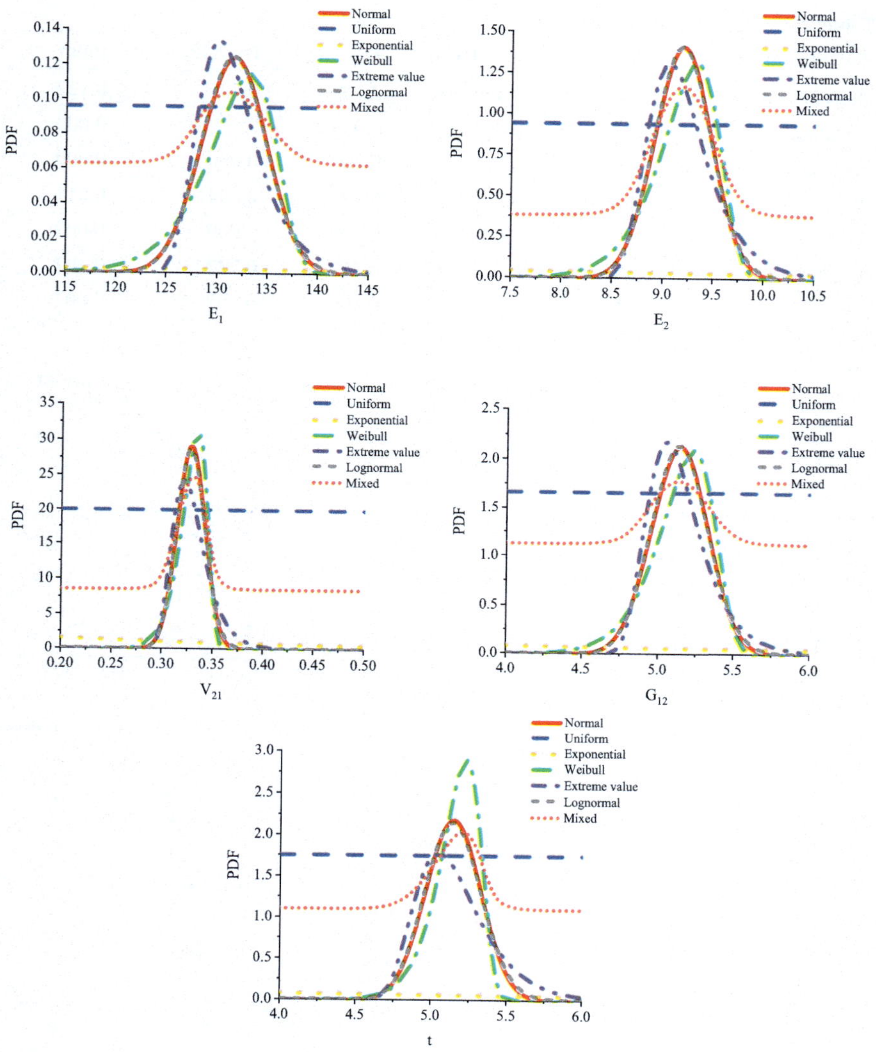

Fig. 2.2 PDF curves of uncertain variables

Table. 2.4 OSL of uncertain variables

Distributions	E_1/GPa	E_2/GPa	V_{21}	G_{12}/GPa	t/mm
Normal	0.596	**0.450**	0.138	0.725	0.122
Uniform	0.137	0.059	0.008	0.339	0.000
Exponential	0.000	0.000	0.000	0.000	0.000
Weibull	0.367	0.262	**0.192**	0.644	**0.217**
Extreme value	0.000	0.000	0.000	0.000	0.000
Lognormal	0.612	0.447	0.119	**0.727**	0.107
Mixed	**0.689**	0.217	0.069	0.557	0.002

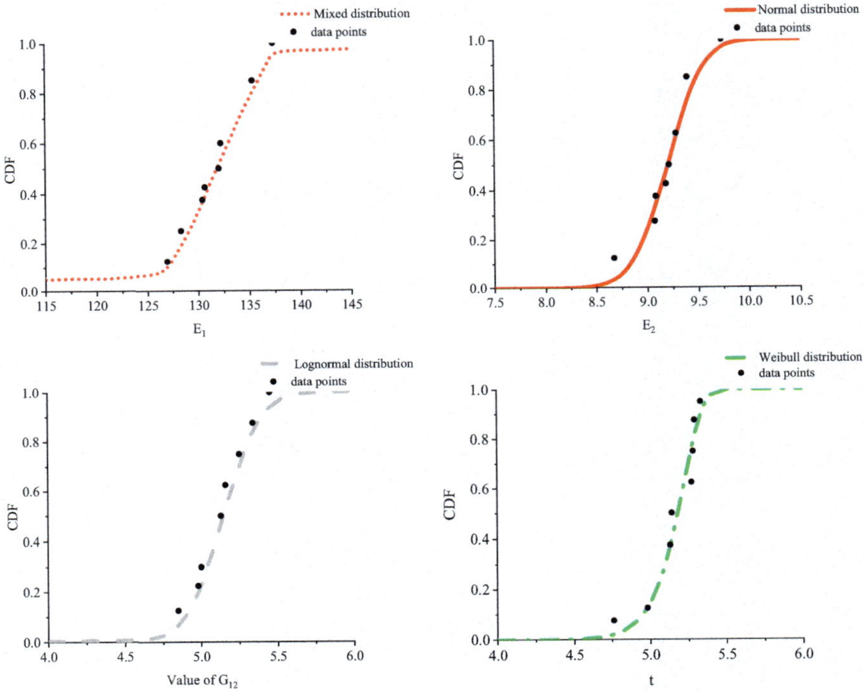

Fig. 2.3 CDF curves of uncertain variables

References

1. Goggin PR. The elastic constants of carbon-fibre composites[J]. J Mater Sci. 1973;8(2):33–44.
2. Elishakoff IE, Li Y, Starnes JH. Non-classical problems in the theory of elastic stability[J]. Cambridge University Press; 2001.

Chapter 3
Aleatory Uncertainty Analysis and Propagation

Under the influence of aleatory uncertainty, the challenge in uncertainty optimisation design lies in efficiently conducting uncertainty analysis and propagation. To address this issue, scholars have proposed various methods, with the MC method being the most commonly used in engineering. However, this method necessitates a significant number of samples to achieve statistical convergence. Moreover, it is noted that the accuracy level of the standard deviation of the response mean can only reach 10^{-N} when the number of sample points is 10^{N+2}. The complexity of employing the MC method is further compounded by the requirement of various types of simulation calculations and a substantial number of uncertainty analyses in ship uncertainty-based design optimisation. Consequently, there is a need to identify a method that maintains the accuracy of the MC method but with lower computational costs.

The PCE method offers a means to quantify the stochastic nature of output from multiple uncertain parameters with varying probability distributions at a reduced computational cost, rendering it more advantageous in engineering applications compared to the MC method. This method is underpinned by a rigorous mathematical theoretical derivation, demonstrating that any element within the space of square productible functions of Gaussian measure can be approximated by an orthogonal polynomial as a basis. By expanding uncertain variables using orthogonal polynomials, the method leverages the properties of these polynomials to derive the stochastic characteristics of random variables from the polynomial coefficients. However, employing the PCE method for ship uncertainty optimisation design presents several challenges:

(1) The statistical regression method, often necessitating large samples, is typically employed to determine the polynomial coefficients in the PCE method, resulting in increased computational demands.
(2) The prevalence of numerous environmental disturbances often gives rise to a multitude of uncertainties, with each additional uncertainty compounding the computational workload.

Z. Liu et al., *Uncertainty-Based Ship Design Optimization*,
https://doi.org/10.1007/978-981-97-2973-9_3

(3) The PCE method does not directly provide the failure probability of constraints. While the most direct approach is to obtain the failure probability by integrating the PDF of the constraint function over the failure domain, it is challenging to directly obtain the PDF using the PCE method.

To address the challenges associated with the PCE method, this chapter proposes several solutions. Firstly, to mitigate the computational demands arising from the statistical regression method in determining polynomial coefficients (problem 1), an improved probabilistic collocation method (IPCM) is introduced. Grounded in the principle of linear irrelevance, IPCM aims to reduce the number of required samples. Secondly, in response to the proliferation of environmental disturbances leading to increased uncertainties (problem 2), a hybrid approach combining the PCE method with the Sobol sensitivity index is proposed. This integration enables the analysis of uncertainty sensitivity and facilitates dimensionality reduction of uncertainty. Thirdly, to tackle the challenge of directly obtaining the PDF of constraints (problem 3), the PCE method is coupled with the maximum entropy method (MEM). This combination allows for the direct determination of constraint PDFs, subsequently enabling the calculation of failure probabilities through integration over the failure domain. Finally, the efficacy of the improved PCE method is demonstrated through numerical simulations and ship optimisation cases, highlighting its practical applicability and effectiveness in addressing the identified challenges.

3.1 PCE

The PCE method originates from the Wiener chaos expansion, which operates on the principle of constructing the model response in the form of a stochastic agent model characterised by high-dimensional orthogonal polynomials. Wiener demonstrated that all random variables following Gaussian distributions with finite variance could be represented as an infinite sum of Hermite polynomials. However, as Hermite polynomials are optimally suited only for Gaussian distributions, recent advancements have extended the classical PCE method to what is known as the gPC method [1]. This extension enables the direct utilisation of various other statistical distributions and their respective orthogonal polynomials. The latest evolution of the PCE method is the arbitrary polynomial chaos (aPC) expansion [2, 3]. The aPC method permits the application of PCE with arbitrary probability distributions, particularly when only limited data with known information is available, and when obtaining solely the higher-order moments of the variables is feasible.

3.1.1 Basic Principle

In the Wiener chaos expansion, all Gaussian-distributed random variables Z in the space of square integrable functions can be represented as follows:

$$
\begin{aligned}
Z(\boldsymbol{\xi}) = \alpha_0 + \sum_{i_1=0}^{n} \alpha_{i_1} \Psi_1(\xi_{i_1}) + \sum_{i_1=0}^{n} \sum_{i_2=0}^{i_1} \alpha_{i_1 i_2} \Psi_2(\xi_{i_1}, \xi_{i_2}) \\
+ \sum_{i_1=0}^{n} \sum_{i_2=0}^{i_1} \sum_{i_3=0}^{i_2} \alpha_{i_1 i_2 i_3} \Psi_3(\xi_{i_1}, \xi_{i_2}, \xi_{i_3}) + \cdots
\end{aligned}
\tag{3.1}
$$

where $\boldsymbol{\xi} = \{\xi_1, \xi_2, \cdots, \xi_n\}$ is a mutually independent n-dimensional random input variable; $Z(\boldsymbol{\xi})$ is a random output variable; α_i are the coefficients of the PCE terms; and $\Psi_i(\boldsymbol{\xi})$ is a multidimensional orthogonal polynomial base with $\boldsymbol{\xi}$ as a variable satisfying:

$$
\int_{\Omega} \Psi_i(\xi) \Psi_j(\xi) \rho(\xi) \mathrm{d}\xi = 0, \quad i \neq j \tag{3.2}
$$

To estimate the uncertainty of the output, Eq. (3.2) is truncated to a certain order. Thus, the d-order PCE of $Z(\boldsymbol{\xi})$ with respect to the n-dimensional random variable $\boldsymbol{\xi}$ is given by:

$$
\begin{aligned}
Z(\boldsymbol{\xi}) \approx \alpha_0 + \sum_{i_1=0}^{n} \alpha_{i_1} \Psi_1(\xi_{i_1}) + \sum_{i_1=0}^{n} \sum_{i_2=0}^{i_1} \alpha_{i_1 i_2} \Psi_2(\xi_{i_1}, \xi_{i_2}) + \cdots \\
+ \sum_{i_1=0}^{n} \sum_{i_2=0}^{i_1} \cdots \sum_{i_d=0}^{i_{d-1}} \alpha_{i_1 i_2 i_3} \Psi_d(\xi_{i_1}, \xi_{i_2}, \cdots, \xi_{i_d})
\end{aligned}
\tag{3.3}
$$

where the orthogonal polynomial basis is:

$$
\Psi_k(\xi_{i_1}, \xi_{i_2}, \cdots, \xi_{i_k}) = (-1)^k (e^{\frac{1}{2}\xi^T \xi}) \frac{\partial^k (e^{-\frac{1}{2}\xi^T \xi})}{\partial \xi_{i_1} \partial \xi_{i_2} \cdots \partial \xi_{i_k}} \tag{3.4}
$$

For convenience, Eq. (3.3) is simplified to:

$$
Z(\boldsymbol{\xi}) \approx \sum_{i=0}^{P-1} \alpha_i \Psi_i(\boldsymbol{\xi}) \tag{3.5}
$$

where $P = \frac{(n+d)!}{n!d!}$.

Table 3.1 Correspondence between distributions of random variables and families of orthogonal polynomials

Variable types	Distribution types	Family of orthogonal polynomials	Range
Continuous	Gaussian distribution	Hermite	$(-\infty, +\infty)$
	Gamma distribution	Laguerre	$[0, +\infty)$
	Beta distribution	Jacobi	$[a, b]$
	Uniform distribution	Legendre	$[a, b]$
Discrete	Poisson distribution	Charlier	$\{0, 1, 2, \cdots\}$
	Binomial distribution	Krawtchouk	$\{0, 1, 2, \cdots, n\}$
	Negative binomial distribution	Meixner	$\{0, 1, 2, \cdots\}$
	Hypergeometric distribution	Hahn	$\{0, 1, 2, \cdots, n\}$

Cameron et al. [4] demonstrated that the Wiener chaos expansion converges for any stochastic process in the ideal scenario. In this scenario, the PDF of the Gaussian distribution aligns with the weighting function of the Hermite multinomial, facilitating rapid convergence of the PCE. However, for other probability distributions like gamma, beta, and uniform distributions, the convergence speed may not be guaranteed if the Hermite multinomial is retained.

Askey et al. [5] introduced orthogonal polynomials tailored for various distributions. Subsequently, Xiu et al. utilised the Askey family to extend the Wiener chaos expansion to encompass prevalent non-Gaussian distributions. Termed the gPC expansion, this approach enables the direct utilisation of non-Gaussian probability distributions in generating polynomial response functions. Such an adaptation has proven effective in enhancing the convergence of the chaos expansion for non-Gaussian random variables. The relationship between the distribution of random variables and the family of orthogonal polynomials is outlined in Table 3.1.

If a one-dimensional orthogonal polynomial base $\varphi_t^{l_t}(\xi_t)$ belongs to the Askey family, and ξ_t is denoted as a one-dimensional orthogonal polynomial base of order l_t of a random variable, and assuming that the n-dimensional random variables ξ are independent of each other, then a multidimensional orthogonal polynomial base out of $\Psi_i(\xi)$ can be decomposed as a product of multiple one-dimensional orthogonal polynomial bases, namely:

$$\Psi_i(\xi_1, \xi_2, \cdots \xi_n) = \prod_{t=1}^{n} \varphi_t^{l_t}(\xi_t), i = 0, 1, \cdots, P-1;\ l_t = 0, 1, \cdots d;\ \sum_{t=1}^{n} l_t \leq d \tag{3.6}$$

3.1.2 Solving for Polynomial Mixed Expansion Coefficients

Once the orthogonal polynomial base of the PCE is established, the subsequent step involves determining the PCE coefficients. Intrusive methods necessitate modifications and adjustments to the model, making them challenging to replicate and not universally applicable. Therefore, this book introduces a widely used non-intrusive method—the regression method.

Let $\{\xi^{(i)}\}$ denote the input sample set of random parameters, with its corresponding output $y = \{y^{(i)}\}0$ being a vector of coefficients whose approximation $\alpha{=}(\alpha_0{,}\alpha_1, \cdots, \alpha_{P-1})$ can be solved by least squares, namely:

$$\hat{\alpha} = \arg\min \sum_{j=1}^{N} \left(y^{(i)} - \sum_{i=0}^{P-1} \alpha_i \Psi_i(\xi^{(i)}) \right) \tag{3.7}$$

where n > P is the number of samples.

Let

$$\Gamma = \begin{bmatrix} \Psi_0(\xi^{(0)}) & \Psi_1(\xi^{(0)}) & \cdots & \Psi_{P-1}(\xi^{(0)}) \\ \Psi_0(\xi^{(1)}) & \Psi_1(\xi^{(1)}) & \cdots & \Psi_{P-1}(\xi^{(1)}) \\ \vdots & \vdots & \vdots & \vdots \\ \Psi_0(\xi^{(n)}) & \Psi_1(\xi^{(n)}) & \cdots & \Psi_{P-1}(\xi^{(n)}) \end{bmatrix} \tag{3.8}$$

Then

$$\hat{\alpha} = (\Gamma^{\mathrm{T}}\Gamma)^{-1}\Gamma^{\mathrm{T}} y \tag{3.9}$$

where $\Gamma^{\mathrm{T}}\Gamma$ is the Fisher matrix.

t is evident that solving the PCE coefficients requires the generation of a series of samples, which can be obtained by simple random sampling, Latin Hypercube sampling (LHS), and so on.

After determining the polynomial expansion coefficients, the mean and standard deviation of the response output, which is the focus of RDO, can be obtained directly. The mean value of the response output is expressed as:

$$E(Z) = \langle \hat{Z}(\varepsilon) \rangle \approx \sum_{i=0}^{P-1} \alpha_i \langle \Psi_i(\varepsilon) \rangle = \alpha_0 \tag{3.10}$$

where $\langle \cdot \rangle$ represents the inner product.

The standard deviation of the response output is expressed as:

$$D(Z) = \langle (Z(\varepsilon) - \langle \hat{Z}(\varepsilon) \rangle)^2 \rangle = \sum_{i=1}^{P-1} \sum_{j=1}^{P-1} \alpha_i \alpha_j \langle \Psi_i(\varepsilon) \Psi_j(\varepsilon) \rangle = \sum_{i=1}^{P-1} \alpha_i^2 \langle \Psi_i(\varepsilon)^2 \rangle \tag{3.11}$$

The regression method necessitates the initial generation of sample points when solving the PCE coefficients. However, the construction and size selection of the sample set currently rely solely on empirical methods, significantly impacting computational efficiency. The quantity of samples utilised determines the number of calculations of output variable values, directly influencing computational costs. Moreover, the distribution of samples also impacts the fitting effectiveness of the PCE, thereby affecting the accuracy of subsequent results. To ensure calculation accuracy, statistical methods employing large samples are often employed to select sample points for regression solving, leading to a substantial increase in computational workload. Therefore, to diminish the requisite sample size, enhance computational efficiency, and reduce computational costs, a novel sampling method is imperative for solving polynomial coefficients.

3.2 IPCM

This section harnesses the orthogonal nature of polynomials and introduces an IPCM derived from the conventional Probabilistic Collocation Method (PCM). Unlike the large-sample statistical regression method based on the principle of linear irrelevance, IPCM offers a more efficient approach by significantly reducing the number of required sample points.

3.2.1 *Basic Principles and Improvements*

The traditional PCM, initially employed in fields such as groundwater seepage simulation and solute transport [6, 7], has seen broader applications in recent years. Its usage now extends to calculating slopes and assessing structural reliability [8, 9], yielding positive outcomes across these domains.

Considering ξ as a one-dimensional variable, the residuals of the truncated PCE with respect to the original model $Z(\xi)$ are

$$R(\alpha, \xi) = Z(\xi) - \hat{Z}(\xi) \tag{3.12}$$

Thus, by ensuring orthogonality between the residuals $R(\alpha_i, \xi)$ and the one-dimensional orthogonal polynomials $\varphi_i(\xi)$, we obtain:

$$\int_{\Omega} R(\alpha, \xi)\varphi_i(\xi)\mathrm{d}\xi = 0 \tag{3.13}$$

Equation (3.13) can be numerically solved using the Gaussian integration method, which results in:

$$\int_{\Omega} R(\alpha, \xi)\varphi_i(\xi)\mathrm{d}\xi \approx \sum_{j=0}^{P-1} w_j R(\alpha, \xi_j)\varphi_i(\xi_j) = 0, \quad i = 0, 1, \cdots, P-1 \tag{3.14}$$

where w_j is the weight, ξ_j is the horizontal coordinate of the Gaussian integration point of the product equation.

For any i and j, $w_j\varphi_i(\xi_j)$ is not always 0; hence, we have:

$$R(\alpha, \xi_j) = 0, \ j = 0, 1, \cdots, P-1 \tag{3.15}$$

As the orthogonal polynomial $\varphi_i(\xi)$ is orthogonal to any polynomial difference $R(\alpha, \xi)$ of not more than dth-order, the Gaussian points represent the zeros of the orthogonal polynomial $\varphi_i(\xi)$ of the (d + 1)th-order order polynomial. Equation (3.15) can be utilised for computing the coefficients of the polynomials using Gaussian integration points.

In summary, for a dth-order PCE of a 1-dimensional random variable, the collocation points correspond to the roots of a (d + 1)th-th order one-dimensional orthogonal polynomial. For instance, the Hermite orthogonal polynomial, being the optimal polynomial for the normal distribution, peaks at the origin. Therefore, the collocation points should be arranged as close to the origin as possible and be symmetrically distributed around it. PCE expansion of odd orders necessitates using the roots of even-ordered Hermite polynomials as collocation points, whereas even-ordered polynomials lack zero roots. Hence, zero roots are included as collocation points for the PCE expansion of odd orders [10]. For example, the roots of Hermite polynomials up to the fourth order are detailed in Table 3.2.

For an n-dimensional random variable, the probabilistic collocations are combinations of the roots of orthogonal polynomials of higher first order in 1 dimension. Thus, the PCM generates all available probabilistic collocations N as:

Table 3.2 Roots of Hermite polynomials

PCE expansion order	Hermite in higher 1 dimension	Roots
1	$\varepsilon^2 - 1$	$0, 1, -1$
2	$\varepsilon^3 - 3\varepsilon$	$0, \sqrt{3}, -\sqrt{3}$
3	$\varepsilon^4 - 6\varepsilon^2 + 4$	$0, \sqrt{3+\sqrt{6}}, \sqrt{3-\sqrt{6}}, -\sqrt{3+\sqrt{6}}, -\sqrt{3-\sqrt{6}}$
4	$\varepsilon^5 - 10\varepsilon^3 + 15\varepsilon$	$0, \sqrt{5+\sqrt{10}}, \sqrt{5-\sqrt{10}}, -\sqrt{5+\sqrt{10}}, -\sqrt{5-\sqrt{10}}$

$$N = (d + 1)^n + \begin{cases} 1, \text{if d is odd} - \text{number} \\ 0, \text{if d is even number} \end{cases} \tag{3.16}$$

Unlike statistical sampling methods, the input samples of PCM are not selected randomly; instead, they follow specific rules. Generally, the number of sample points should exceed the number of coefficients to be determined. Isukapalli [11] suggested that the number of collocations should be twice the number of coefficients to be determined. Subsequently, the coefficients are determined using the regression method. However, Jiang et al. [12] discovered that sometimes, despite having more than eight times the number of collocations compared to the coefficients to be determined, calculation accuracy could not be guaranteed. Additionally, increasing collocation points often reduces calculation efficiency, sometimes even surpassing that of the MC method.

From the perspective of linear algebra, as outlined in Eq. (3.9), ensuring that the coefficients can be computed and that they represent unique solutions requires the coefficient matrix to be of full rank, with a rank equal to the number of undetermined coefficients. Therefore, it is unnecessary to employ all N probabilistic collocations. Instead, selecting P probabilistic collocations from the available N ensures that the coefficient matrix achieves full rank, thus obtaining a unique solution for the coefficients.

In Eq. (3.9), it is evident that each column of the coefficient matrix corresponds to the same multidimensional polynomial. Consequently, each row shares an identical format with the preceding one, differing solely in the substitution of the collocation points. Given the symmetry of the collocation points around the origin and the inclusion of even terms, a linear correlation emerges between the two row vectors within the coefficient matrix. If the row vectors of the newly introduced collocation point exhibit linear dependency on those of the selected collocation point, it signifies that the additional collocation point fails to offer further effective information for determining the coefficients. Consequently, the collocation point is deemed unsuitable, prompting its rejection and the selection of the subsequent candidate point for continued evaluative analysis. Conversely, if the row vectors of the newly introduced collocation point are not linearly correlated with those of the selected collocation point, it indicates that the newly introduced probabilistic collocation point can indeed provide valuable information for determining the coefficients of the polynomial. Thus, retaining such a collocation point becomes imperative.

Hence, this book introduces an IPCM grounded in the principle of linear irrelevance, ensuring that the row vectors of the polynomial matrix are linearly independent, thereby constituting a matrix of full row rank. The flowchart illustrating the linear-independent PCM based on linear independence is depicted in Fig. 3.1.

When dealing with a large number of random variables and a high order of PCE expansion, the computational burden of identifying linearly independent collocations becomes substantial. This process entails multiple iterations of constructing the coefficient matrix and determining its rank, which is time-intensive. Typically, for a specified number of random variables and a set expansion order of a random

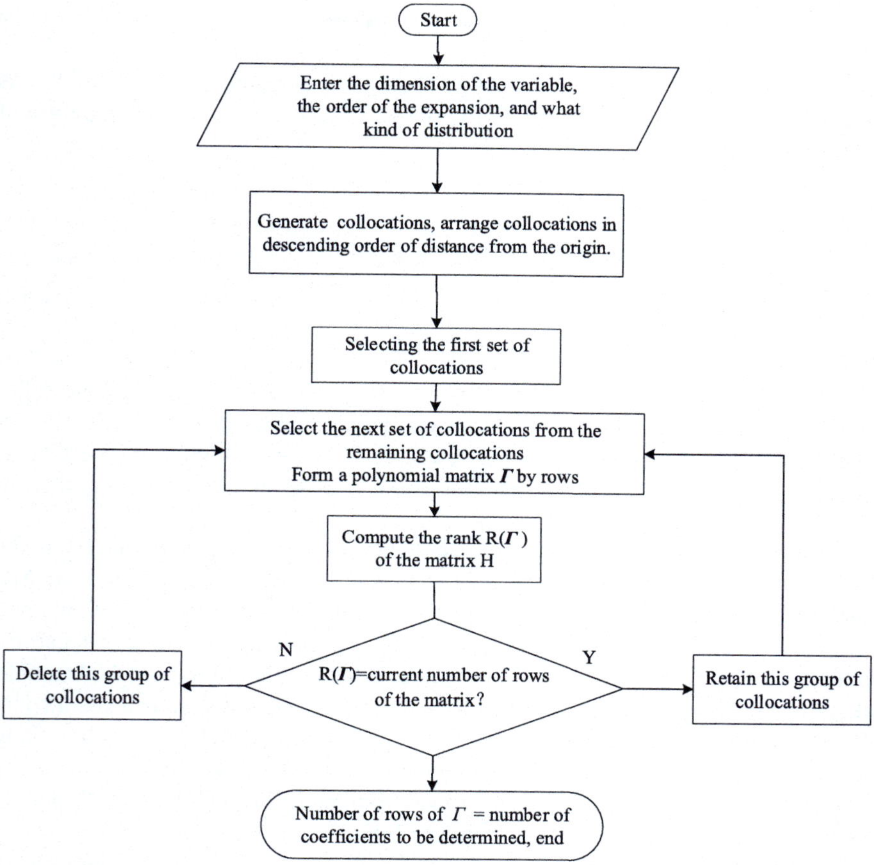

Fig. 3.1 Flowchart of PCM based on linear-independent

polynomial, it is feasible to pre-search for and store the necessary linearly independent collocations. This enables direct access to them without the need for repeated searches.

3.2.2 Numerical Examples

1. **2-Dimensional 3rd order polynomial functions**

Consider the 2-dimensional 3rd order nonlinear function described by Eq. (3.17), wherein the random variables x_1 and x_2 adhere to normal distributions with mean values μ of 10 and 9.9, respectively, and standard deviations δ of 5. It is noteworthy that x_1 and x_2 are independent of each other. This setup can be represented as:

$$Y_1 = g(x_1, x_2) = x_1^3 + x_1^2 x_2 + x_2^3 - 18 \tag{3.17}$$

Transforming the random variables x_1 and x_2 into standard normal distributions as $\varepsilon = (x - \mu)/\delta$, substituting them into Eq. (3.17), and rearranging into the form of a Hermite polynomial, we obtain:

$$\begin{aligned} Y_1 &= g(\varepsilon_1, \varepsilon_2) \\ &= 4682.299 + 2865\varepsilon_1 + 2470.15\varepsilon_2 + 997.5(\varepsilon_1^2 - 1) + 742.5(\varepsilon_2^2 - 1) \\ &\quad + 500\varepsilon_1\varepsilon_2 + 125(\varepsilon_1^3 - 3\varepsilon_1) + 125(\varepsilon_2^3 - 3\varepsilon_2) + 125(\varepsilon_1^2 - 1)\varepsilon_2 \end{aligned} \tag{3.18}$$

The function Y_1 is expanded up to the 2nd, 3rd, and 4th orders using the PCE method, employing 6, 10, and 15 expansion terms, respectively. To select the collocation points ε_1 and ε_2, three methods are employed: PCM, IPCM, and MC method. In each method, the number of collocations is set to twice the number of coefficients to be determined. The distribution of collocations for the 2nd, 3rd, and 4th orders in 2 dimensions is depicted in Fig. 3.2.

From Fig. 3.2, it is evident that the distribution of collocations produced by the PCM exhibits a certain level of regularity and is more uniform compared to the MC method. Particularly, when the number of collocations is limited, the MC method may exhibit insufficient collocations in certain directions or even lack collocations entirely. In the IPCM, adhering to the principle of linear irrelevance, a reduced number of collocations (6, 10, 15) are selected from the (9, 25, 25) collocations generated by the PCM for the 2nd, 3rd, and 4th order expansions, respectively. This reduction in the number of samples is noteworthy compared to the other two methods.

A comparison of the coefficients obtained from the IPCM, PCM, and MC method for various expansion orders is presented in Table 3.3.

Upon comparing with the exact value of Eq. (3.18), for the function represented by Y_1, when the expansion order is 2nd, the polynomial coefficients acquired through both collocation methods lack precision, with one coefficient Y_1 differing from the exact value. However, when the expansion order is elevated to the 3rd, aligning with the real order of Y_1, the coefficients computed by the PCE method accurately reflect those of the function Y_1. Notably, the polynomial coefficients derived from both collocation methods closely match the exact values, with negligible error. In scenarios involving 2nd and 3rd order expansions, the number of collocation points obtained via the IPCM is fewer than that acquired through the PCM. Despite this, IPCM demonstrates comparable accuracy to PCM. This underscores the advantage of IPCM in predicting the mean and standard deviation of the response. Specifically, IPCM achieves this with fewer sample points compared to PCM, thereby reducing computational costs while maintaining accuracy.

Upon reaching the 4th order of expansion, the coefficient values of the 4th order term obtained through both collocation methods closely approach 0. It can be inferred that the 4th order term does not significantly enhance the accuracy of this numerical

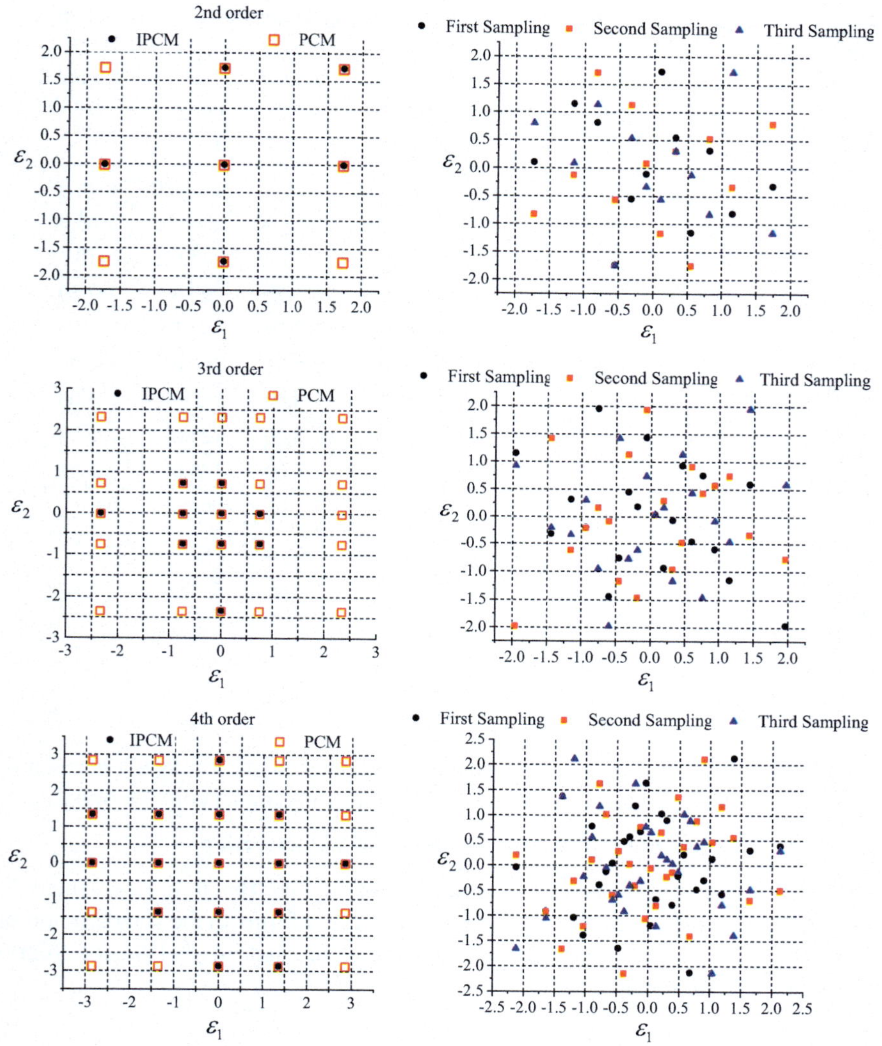

Fig. 3.2 Distribution of 2-dimensional 2nd order, 2-dimensional 3rd order, and 2-dimensional 4th order collocations (Left column is 2 probabilistic collocations methods, right column is MC method with 3 samples)

example. Therefore, it can be concluded that the optimal expansion order for this scenario is the 3rd order.

Hence, it is not universally true that increasing the expansion order leads to greater accuracy. Each problem in PCE expansion has an optimal expansion order associated

Table 3.3 Comparison of coefficients obtained by IPCM, PCM and MC methods with different expansion orders

Coefficient	Order 2		Order 3		Order 4		Accuracy value (MC)
	IPCM	PCM	IPCM	PCM	IPCM	PCM	
	6	9	10	25	15	25	
b_0	4628.30	4628.30	4628.30	4628.30	4628.30	4628.30	4628.30
b_1	2864.99	2864.99	2865.00	2865.00	2865.00	2865.00	2865.00
b_2	2345.15	2345.15	2470.15	2470.15	2470.15	2470.15	2470.15
b_3	997.50	997.50	997.50	997.50	997.50	997.50	997.50
b_4	742.50	742.50	742.50	742.50	742.50	742.50	742.50
b_5	**716.50**	**526.36**	500.01	500.01	500.01	500.01	500.00
b_6			125.00	125.00	125.00	125.00	125.00
b_7			125.00	125.00	125.00	125.00	125.00
b_8			125.00	125.00	125.00	125.00	125.00
b_9			0.01	0	0	0	0
b_{10}					0	0	
b_{11}					0	0	
b_{12}					0	0	
b_{13}					0	0	
b_{14}					0	0	

with it. Therefore, indiscriminately raising the expansion order may not necessarily yield more precise results but could result in significantly higher computational costs.

2. **Multidimensional polynomial functions of higher order**

Consider the multidimensional high-order polynomial function represented by Eq. (3.19). The input variable $X_i (i = 1, 2, \cdots, n)$ conforms to the standard normal distribution. For this analysis, we set n = 3, indicating that Eq. (3.19) is of order 6. Consequently, we obtain:

$$Y_2 = \frac{1}{2^n} \prod_{i=1}^{n} (3X_i^2 + 1) \tag{3.19}$$

Expansions from the 2nd to the 6th order are applied to Y_2, with the number of expansion terms set at 10, 20, 35, 56, and 84, respectively. The input variables of Eq. (3.19) are chosen using both the IPCM and PCM methods, as illustrated in Fig. 3.3.

In the case of IPCM, (10, 20, 35, 56, 84) collocations are selected from the (27, 125, 125, 343, 343) collocations generated by PCM following the principle of linear irrelevance. The sieve-reduced collocations are depicted in Fig. 3.3. The distribution of these collocations exhibits a discernible pattern, with points of the same colour equidistant from the origin forming a ladder-like arrangement around it. Additionally,

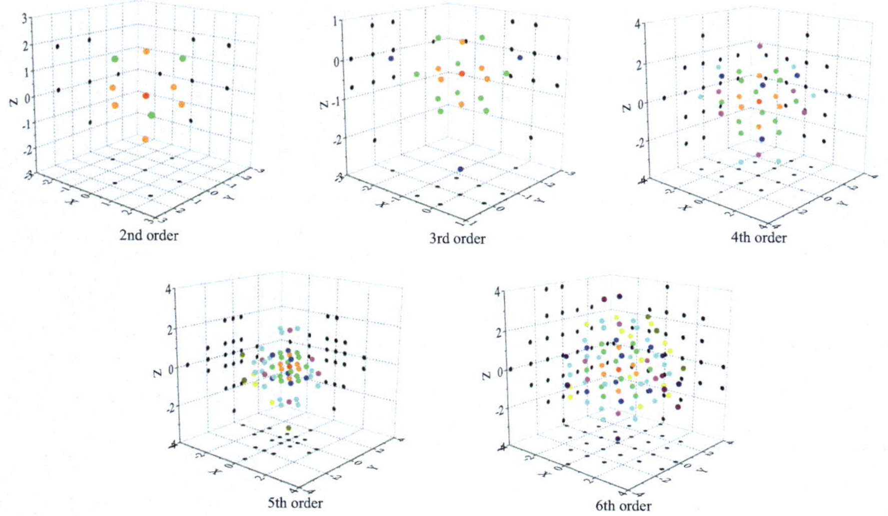

Fig. 3.3 Selection of IPCM points under 3-dimensional problem

the black points represent the projection of the matching points onto each surface, with a similar and uniform distribution of points across each quadrant.

In contrast to Eq. (3.18), the Y_2 function in Eq. (3.19) cannot be reformulated into the form of Hermite polynomials. Consequently, it is impractical to assess the accuracy of IPCM by comparing the errors between the solved polynomial coefficients and the true coefficients. Therefore, this study adopts the results obtained from the MC method (with 10^5 sample points) as the reference value. It considers the PDF curve of the output function Y_2 and compares the PDF curves generated by PCM and IPCM against it. The outcomes are illustrated in Fig. 3.4.

In Fig. 3.4, concerning the function Y_2, a notable observation is made: when the expansion order is 2nd, both the PDF curves of PCM and IPCM exhibit substantial errors across the entire range of Y_2 compared to the MC method. Moreover, the accuracy of PCM exceeds that of IPCM before the peak. Conversely, at the 3rd expansion order, the PDF of PCM demonstrates high accuracy near the peak, albeit with increasing errors post-peak. Meanwhile, IPCM shows decreased errors compared to the 2nd order expansion, notably improving accuracy before the peak. As the unfolding order $d \geq 4$ increases, IPCM's accuracy significantly enhances throughout the range of Y_2, aligning closely with PCM. Notably, the PDF curves of both methods converge towards the results of the MC method, with some discrepancies primarily evident in the tail region with low probability. At this juncture, IPCM requires fewer sample points (35, 56, 84) compared to PCM (125, 343, 343), and even fewer than the MC method (10^5). This underscores IPCM's capability to substantially reduce computational workload while ensuring accuracy.

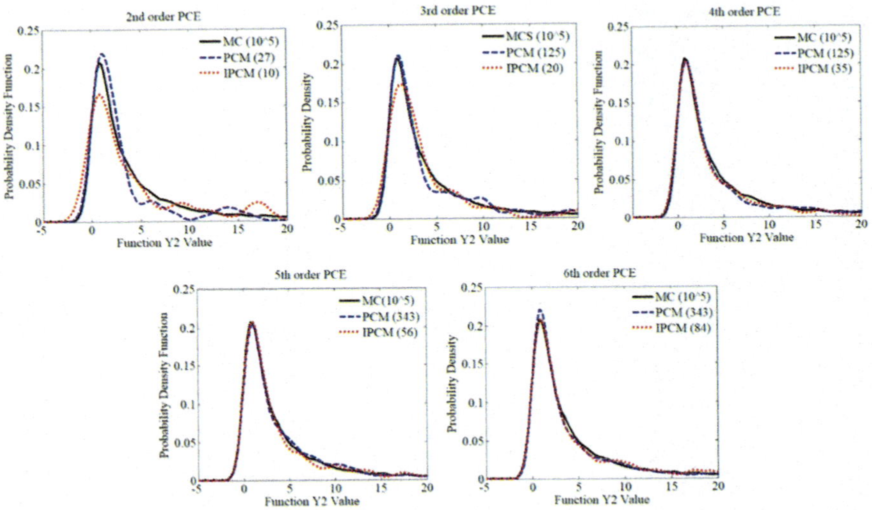

Fig. 3.4 Functions comparison of PDF curves of MC\PCM\IPCM method at different expansion orders

3. **Ishigami function**

Consider the Ishigami function, a nonlinear non-monotonic function expressed as follows:

$$Y_3 = \sin X_1 + a \sin^2 X_2 + bX_3^4 \sin X_1 \tag{3.20}$$

where the input variable $X_i (i = 1, 2, 3)$ is uniformly distributed in $[-\pi, \pi]$.

Here, a = 7 and b = 0.1. Due to the nonlinearity of Y_3, determining the appropriate expansion order for the PCE method is challenging. Hence, the function Y_3 is expanded to the 3rd, 5th, 7th, and 9th orders, with 20, 56, 120, and 220 expansion terms, respectively. The mean and standard deviation of the response Y_3 are obtained using PCM and IPCM for selecting the points of X_1 and X_2, respectively. A comparison with the results of the MC method is presented in Tables 3.4 and 3.5, showcasing the mean and standard deviation of the response Y_3. Furthermore, the variations of these two indices with the unfolding order are illustrated in Fig. 3.5.

Table 3.4 Ishigami function-means and standard deviations calculated by PCM method with different expansion orders

System of fractions	PCM				MC
	$d = 3$	$d = 5$	$d = 7$	$d = 9$	
Distribution points	125	343	729	1331	10^5
Means	2.812	3.279	3.472	3.429	3.485
Standard deviation	3.014	3.536	3.727	3.699	3.691

Table 3.5 Ishigami function-means and standard deviations calculated by IPCM method with different expansion orders

System of fractions	PCM				MC
	$d = 3$	$d = 5$	$d = 7$	$d = 9$	
Distribution points	125	343	729	1331	10^5
Means	2.812	3.279	3.472	3.429	3.485
Standard deviation	3.014	3.536	3.727	3.699	3.691
System of fractions	IPCM				MC
	$d = 3$	$d = 5$	$d = 7$	$d = 9$	
Distribution points	20	56	120	220	10^5
Means	2.676	3.185	3.450	3.435	3.485
Standard deviation	2.711	3.421	3.728	3.725	3.691

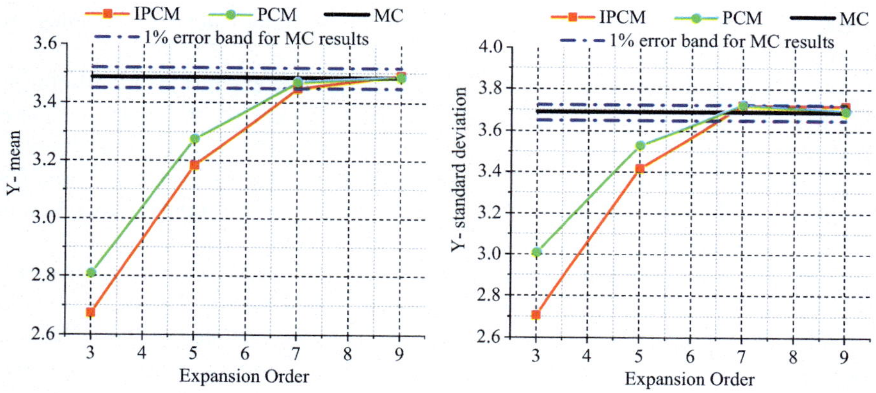

Fig. 3.5 Curve of mean and standard deviation of Y_3 as a function of the order of polynomial expansion

As depicted in Tables 3.4 and 3.5, and Fig. 3.5, the mean and standard deviation errors of both PCM and IPCM computations are more pronounced at lower expansion orders (d = 3, 5). With increasing order, there is a tendency for the mean and standard deviation to converge, approaching the results obtained through the MC method. Specifically, at d = 7, the mean and standard deviation obtained by both methods fall within the 1% error margin of the MC method, indicating that the 7th order polynomials adequately capture the uncertainty of the Y_3. Moving to d = 9, there is a slight improvement in the accuracy of the mean and standard deviation compared to d = 7, albeit with a sample point count nearly double that of d = 7.

At equivalent unfolding orders, while the PCMs exhibit slightly higher accuracy compared to the IPCMs, the IPCM computations necessitate significantly fewer sample points than those of PCM. With the increase in unfolding order, PCM requires 1331 sample points at d = 9, whereas IPCM only requires 220. Consequently, IPCM ensures calculation accuracy with a reduced number of sample points, highlighting its pronounced advantage in efficiency.

4. **Limit state function**

The PCM selects a greater number of sample points compared to the coefficients to be determined, whereas the IPCM ensures accuracy by selecting a subset of sample points (equivalent to the number of coefficients to be determined) from the matching points generated by the PCM for calculation. To delve deeper into this phenomenon, this example examines the relationship between the number of matching points and the matrix rank. Consider a commonly used nonlinear non-monotonic function in reliability analysis, given by:

$$g(x) = x_1 - 10^4\left(\frac{x_2(x_4x_5)^{1.71}}{x_3} + \frac{(1 - x_2)(x_4x_5)^{1.188}}{x_6}\right) \tag{3.21}$$

where the fundamental parameters of the six non-standard normal variables are outlined in Table 3.6.

As $g(x)$ is a nonlinear function, determining the suitable expansion order for fitting this function using the PCE method is not feasible. Therefore, the function $g(x)$ is expanded to the 2nd, 3rd, and 4th order, with the number of expansion terms set at 28, 84, and 210, respectively. Collocations are then selected for the six variables by PCM and IPCM, respectively. Subsequently, the mean and standard deviation of $g(x)$ are calculated concurrently.

Additionally, the mean and standard deviation of g() are computed using the first N collocations selected from the 729 collocations (d = 2nd order), 15,625 collocations (d = 3rd order), and 15,625 collocations (d = 4th order) generated by the PCM. The number of collocations N is set to 2, 3, 4, and 10 times the number of coefficients to be determined, respectively. The results of PCM (with the number of collocations N = 2P, 3P, 4P, 10P, and the full rank, where P is the number of coefficients of the expansion term) and IPCM calculations, along with their error percentages compared to the results of the MC method, are presented in Tables 3.7 and 3.8.

Tables 3.7 and 3.8 present a comparison of the mean and standard deviation of $g(x)$ for different numbers of collocations. Using the results of the MC method as a reference, it is observed that with the increase in the order of expansion, the mean

Table 3.6 Table of basic parameters for non-standard normal variables

Random variable	Distribution type	Means	Standard deviation
x_1	Lognormal distribution	1.044	0.3132
x_2	Normal distribution	0.7	0.07
x_3	Lognormal distribution	0.2391	0.09564
x_4	Lognormal distribution	1.011	0.15165
x_5	Normal distribution	0.0005	0.00008
x_6	Lognormal distribution	1.802	0.7208

Table 3.7 Comparison of mean values of PCE with different number of match points

Projects	PCE						MC (10^5)
	$d = 2$		$d = 3$		$d = 4$		
	Mean values	Errors/%	Mean values	Errors/%	Mean values	/%	
N = 2P	0.689	5.1					0.726
N = 3P	0.688	5.2	0.327	55.1	0.946	30.3	
N = 4P	0.687	5.4	0.544	22.1	0.719	0.9	
N = 10P	0.689	5.1	0.737	21.5	0.732	0.8	
Full rank	0.688	5.2	0.724	0.3	0.727	0.1	
IPCM	0.688	5.2	0.723	0.4	0.728	0.3	

Table 3.8 Comparison of standard deviation of PCE at different number of collocations

Projects	PCE 法						MC (10^5)
	$d = 2$		$d = 3$		$d = 4$		
	Standard deviation	Errors/%	Standard deviation	Errors/%	Standard deviation	Errors/%	
N = 2P	0.389	13.4					0.343
N = 3P	0.385	12.2	0.512	49.2	0.611	78.1	
N = 4P	0.383	11.7	0.492	43.4	0.348	1.5	
N = 10P	0.379	9.6	0.371	8.2	0.347	1.2	
Full rank	0.372	8.5	0.361	5.2	0.341	0.6	
IPCM	0.377	9.9	0.365	6.4	0.346	0.8	

and standard deviation calculated by the PCE method approach those of the MC method, leading to a gradual decrease in error, with the 4th order yielding the most accurate results. Regarding the regression method, for the 2nd order expansion, the increase in the number of collocations has minimal impact on the accuracy of the mean and standard deviation. However, for the 3rd order expansion, increasing the number of collocations to $10P$ reduces the accuracy of the calculation. Conversely, for the 4th order expansion, the accuracy of the calculation improves as the number of collocations increases to $4P$.

To elucidate the relationship between computational accuracy and the number of collocations, Fig. 3.6 illustrates the correlation between the rank of the coefficient matrix and the number of matching points for different expansion orders. It is evident that the coefficient matrix may not achieve full rank when the number of collocations surpasses the number of coefficients to be determined. This circumstance results in significant errors in the calculations, as evidenced in Tables 3.7 and 3.8. For the 2nd order expansion, the coefficient matrix attains full rank when the number of collocations exceeds 52. Subsequently, increasing the number of collocations does not enhance the accuracy of the calculation. Specifically, when $N = 10P = 280$, the

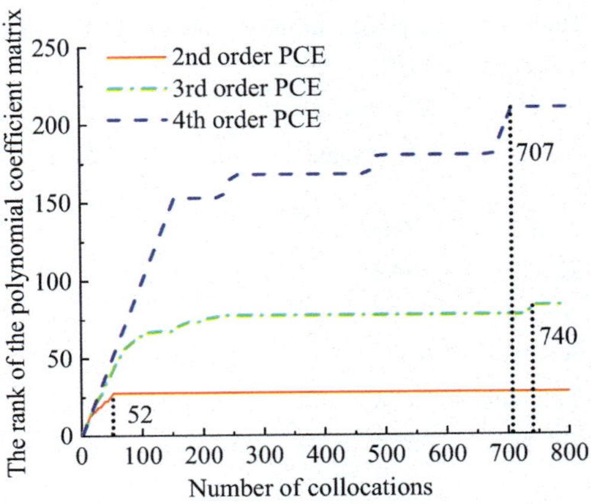

Fig. 3.6 Relationship between the rank of the coefficient matrix and the number of collocation points

relative errors of the mean and standard deviation for the 2nd order expansion are 5.1 and 9.6%, respectively. Conversely, with $N = 52$, where the coefficient matrix already achieves full rank, the relative errors are 5.2% and 10.5%. Despite the former having over five times the number of collocations compared to the latter, there is no significant improvement in computational accuracy. Similarly, for the 3rd order expansion, the matrix reaches full rank when the number of collocations exceeds 740, leading to improved calculation accuracy at $N = 10P = 840$. Conversely, when $N = 2P, 3P, 4P$ (the number of collocations is less than 740), the coefficient matrices fail to achieve full rank, resulting in large errors or even unsolvable solutions. This trend holds true for the 4th order expansion as well. Furthermore, the rank of the coefficient matrix does not increase monotonically with the number of coordination points. For instance, in the case of the 4th order expansion, the rank remains constant at 153 from the 153rd to the 233rd collocations. This implies linear correlation among these collocations, rendering it sufficient to select only one group while eliminating the rest.

When employing the regression method, the minimum number of collocations necessary to ensure the full rank of the coefficient matrix is 52 for the 2nd order, 740 for the 3rd order, and 707 for the 4th order expansions. Conversely, when utilising IPCM, the number of collocations resulting in a full-rank coefficient matrix is 28 for the 2nd order, 84 for the 3rd order, and 220 for the 4th order expansions. This illustrates IPCM's capability to significantly reduce the number of collocations required for computation, thereby lowering the computational cost. While the accuracy of the former calculation slightly surpasses the latter due to the latter's utilisation of only a fraction of the allocation points of the former, both methods produce results that adequately meet the accuracy requirement compared to the exact solution. Moreover, IPCM demonstrates enhanced computational efficiency.

In summary, when choosing allocation points for PCM and resolving the polynomial coefficients using the regression method, it is imperative to ensure that the selected probability collocations satisfy the criterion whereby the rank of the coefficient matrix equals the number of coefficients to be determined. This ensures the accuracy of the PCE method. However, in instances involving high dimensions and ranks, this method becomes computationally intensive. Conversely, IPCM, which operates on the principle of linear irrelevance, mandates selecting the same number of collocations as the coefficients to be determined. This approach significantly enhances computational efficiency. Consequently, IPCM presents notable advantages in complex uncertainty analysis and simulation calculations.

3.3 Global Sensitivity Analysis Based on PCE Method

To authentically depict the performance of a design solution amidst varying degrees of uncertainty, designers aim to incorporate an expanding array of uncertain quantities during the uncertainty design optimisation process. By integrating more uncertainty factors, designers seek to derive solutions that are both robust and reliable, mimicking real-world operational scenarios. However, the inclusion of numerous uncertain quantities gives rise to high-dimensional problems, presenting significant challenges in uncertainty optimisation design and leading to inefficient optimisation processes. As illustrated in Fig. 3.7, the minimum number of samples required to construct a PCE model escalates with the augmentation of input variables. This trend becomes more pronounced in slightly higher-order expansions, consequently amplifying the computational burden associated with modelling.

Hence, within the realm of uncertainty-optimised design, dimensionality reduction techniques have emerged as viable solutions to mitigate this challenge, aiming to strike a balance between fostering robust and dependable solutions and minimising computational overheads. Sensitivity analysis serves as a means to qualitatively or

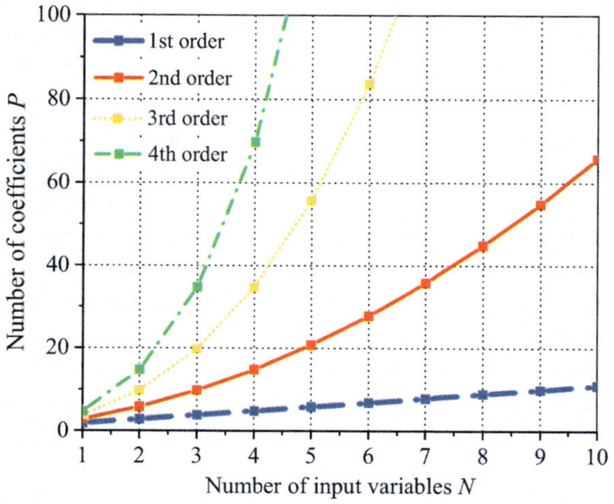

Fig. 3.7 Minimum number of samples (number of coefficients) required for the PCE model versus input variables

quantitatively assess the influence of model parameter uncertainty on model outputs, thus facilitating the prioritisation of uncertainties based on their impact and enabling dimensionality reduction. Sensitivity analysis methods are categorised into local sensitivity analysis and global sensitivity analysis. The latter examines the overarching impact of model uncertainty inputs on model outcomes across their permissible range of variation, making it the prevalent approach currently utilised.

In the realm of global sensitivity analysis methods, several approaches prevail, including the multiple regression method, Morris' method, Fourier magnitude sensitivity test, and Sobol's global sensitivity method based on variance decomposition. This book centres on Sobol's global sensitivity method based on variance decomposition [13]. This method dissects the variance of the output function into individual parameters and the bias variance among these parameters. It utilises the ratio of the bias variance to the output variance as an indicator of each parameter's contribution and the interaction among parameters to the output variance. Sobol's method typically entails sampling the integral of the function using the MC method to calculate the variance. To achieve statistical convergence, this method necessitates a substantial number of samples; otherwise, MC error occurs when the sample points are infinite. Conversely, employing the PCE method for sensitivity analysis of uncertainties yields both the desired stochastic statistics and the global sensitivity index of Sobol directly. Consequently, once the polynomials of the output variables are established, resolving the sensitivity index becomes a mere post-processing task of the polynomial coefficients.

Deman et al. [14] were pioneers in extending the polynomial chaos approach as an alternative to the conventional Sobol MC method for computing Sobol's global sensitivity metrics. Within the context of stochastic modelling of subsurface flow, the PCE method has been extensively employed and demonstrated to be a comprehensive and dependable tool for conducting sensitivity analysis at a low computational expense. Deman et al. integrated the computation of Sobol's sensitivity metrics with the sparse PCE method to effectively analyse the impact of uncertainty in input parameters on the model output of groundwater throughout its life cycle.

Fajraoui et al. [15] and Younes et al. [16] applied PCE-based global sensitivity analysis to a fine-grid numerical model of flow in a non-homogeneous porous medium. They addressed uncertain flow boundary conditions concerning liquid conductivity and dispersion rate of solute concentration at specific observation points. Sochala et al. [17] utilised the PCE method to assess the effect of soil parameter uncertainty on three distinct subsurface unsaturated flow physical models. Their findings demonstrated that the PCE method offers greater efficiency compared to the classical MC method and can accurately capture output variations at a reduced computational cost. In the simulation of isotope transport in aquifers, Valentina et al. [18] analysed the statistical moments of peak solute concentrations at designated locations. They explored the relationship between dispersion and partition coefficients, which are influenced by the conductivity field in inhomogeneous media. Sobol' indices obtained from PCE models of various orders indicated that lower-order models could yield reliable metrics while significantly mitigating computational burdens. Formaggia et al. [19] employed a sensitivity index based on PCE to investigate the influence of uncertainty in hydrogeological variables on sedimentation processes at the watershed scale.

To mitigate the escalation of computational costs due to an excessive selection of uncertainties, this section employs the combination of the PCE method with Sobol's sensitivity index for conducting sensitivity analysis on the uncertainties. This approach facilitates the downsizing of uncertainties, ensuring a more manageable computational workload.

3.3.1 Fundamentals and Expansion

The variance decomposition method proposed by Sobol breaks down the variance of the model output into the sum of conditional variances of each input variable and the interaction term of the input variables. Subsequently, the ratio of the conditional variance to the variance of the model output serves as the sensitivity index. Assuming $x = (x_1, x_2, \cdots, x_n)$ represents an n-dimensional random input variable with a PDF of $f_{\boldsymbol{x}}(x)$, $\boldsymbol{x} \in I$. The variables are independent, and their respective marginal PDFs are $f_{x_i}(x_i)$, $i = 1, 2, \cdots, n$. For a square integrable function $Y(x)$, it can be expanded as follows:

$$\begin{aligned} Y(\boldsymbol{x}) &= Y_0 + \sum_{i=1}^{n} Y_i(x_i) + \sum_{i<j} Y_{ij}(x_i, x_j) + \cdots + Y_{12\cdots n}(x_1, x_2, \cdots, x_n) \\ &= Y_0 + \sum_{s=1}^{n} \sum_{i_1<i_2<\cdots<i_s}^{n} Y_{i_1 i_2 \cdots i_s}(x_{i_1}, x_{i_2}, \cdots x_{i_s}) \\ &= Y_0 + \sum Y_{\boldsymbol{u}}(x_{\boldsymbol{u}}) \end{aligned} \tag{3.22}$$

where $1 \leq i_1 \leq i_2 \leq \cdots \leq i_s \leq n, u = (i_1, i_2, \cdots, i_s)$, $x_{\boldsymbol{u}}$ denotes the subvector of x. The integral of the expansion term for any independent parameter is 0, and the decomposition of Eq. (3.22) is unique, i.e.:

$$\int Y_{\boldsymbol{u}}(x_{\boldsymbol{u}}) f_t(x_t) \mathrm{d}x_t = 0, \quad t \in u \tag{3.23}$$

The orthogonality property is derived from Eq. (3.23), as follows:

$$E[Y_{\boldsymbol{u}}(x_{\boldsymbol{u}}) Y_V(x_V)] = 0, \quad u \neq v \tag{3.24}$$

Integrating both sides of Eq. (3.22) by squaring, we get:

$$
\begin{aligned}
&\int Y^2(x)f_x(x)\mathrm{d}x - Y_0^2 \\
&= \sum_{i=1}^{n} \int Y_i^2(x_i)f_{x_i}(x_i)\mathrm{d}x_i + \sum_{i<j} \int Y_{ij}^2(x_i, x_j)f_{x_i}(x_i)f_{x_j}(x_j)\mathrm{d}x_i\mathrm{d}x_j + \cdots \\
&+ \int Y_{12\cdots n}^2(x_1, x_2, \cdots, x_n)f_{x_1}(x_1)f_{x_2}(x_2)\cdots f_{x_n}(x_n)\mathrm{d}x_1\mathrm{d}x_2\cdots\mathrm{d}x_n
\end{aligned}
\tag{3.25}
$$

where the left side of the equation is the variance V of the output of $Y(x)$; whereas the right side is the conditional variance of the corresponding output of each variable.

Let the 1st order conditional variance be:

$$
V_i = \int Y_i^2(x_i)f_{x_i}(x_i)\mathrm{d}x_i,
$$

the 2nd order conditional variance be:

$$
V_{ij} = \int Y_{ij}^2(x_i, x_j)f_{x_i}(x_i)f_{x_j}(x_j)\mathrm{d}x_i\mathrm{d}x_j
$$

and so on, with:

$$
V = \sum_{i=1}^{n} V_i + \sum_{i<j} V_{ij} + \cdots + V_{12\cdots n} = \sum_{s=1}^{n} \sum_{i_1<i_2<\cdots<i_s}^{n} V_{i_1 i_2 \cdots i_s} = \sum V_{\boldsymbol{u}} \tag{3.26}
$$

According to the definition of Sobol' sensitivity index, it can be inferred that:

$$
SI_{\boldsymbol{u}} = \frac{V_{\boldsymbol{u}}}{V} = \frac{E[Y_{\boldsymbol{u}}^2(x_{\boldsymbol{u}})]}{V} \tag{3.27}
$$

And $\sum SI_{\boldsymbol{u}} = 1$.

The first-order SI, which solely considers a single variable x and portrays its individual impact on the model output, is also referred to as the main effect sensitivity index (MSI). Similarly, the second-order sensitivity index SI_{ij} accounts for the influence of the interaction between two variables, x_i and x_j, on the model output. By extension, other higher-order SI can be derived. It is evident that considering all higher-order effects would necessitate a substantial number of SI to be calculated. Therefore, in practical applications, all higher-order effects are often measured by the total effect sensitivity index TSI, which characterises the sum of the MSI of the variable x_i, and the higher-order SI that includes x_i.

As outlined in Sect. 3.1, $l_t = 0, 1, \cdots d$ and $\sum_{t=1}^{n} l_t \leq d$. Definition ϑ is a set composed of $l = (l_1, l_2, \cdots, l_n)$, i.e.:

$$
\vartheta_{\boldsymbol{u}} = \left\{ l : \left| \begin{matrix} l_t > 0, & \forall t = 1, 2, \cdots, n, & t \in u \\ l_t = 0, & \forall t = 1, 2, \cdots, n, & t \notin u \end{matrix} \right. \right\} \tag{3.28}
$$

Comparing Eqs. (3.1) and (3.22), we obtain:

$$Y_u(\boldsymbol{x}_u) = \sum_{l \in \vartheta_u} \alpha_l \Psi_l(\boldsymbol{\varepsilon}) \tag{3.29}$$

Similarly, considering Eqs. (3.1) and (3.27), we get:

$$SI_u = \frac{\sum_{l \in \vartheta_u} \alpha_l^2 . E\left(\Psi_l^2\right)}{D_{PC}} \tag{3.30}$$

Hence, utilising the coefficients and bases of PCE enables the derivation of various sensitivity indices.

3.3.2 Numerical Example

Below, several typical standard functions are selected, and their Sobol' sensitivity indices are computed and compared with the analytical solutions.

1. **Polynomial functions**

First, let us consider the following simple polynomial function:

$$SI_{i_1 i_2 \cdots i_s} = \frac{5^{-s}}{(6/5)^n - 1} \quad (i = 1, 2, \cdots, n) \tag{3.31}$$

For the analytical solution of its Sobol' metric, please refer to paper [20].

Here we take $n = 3$, i.e., the polynomial is of order 6, in which case the Sobol' metric is:

$$\begin{gathered} SI_1 = SI_2 = SI_3 = 25/91 = 0.2747 \\ SI_{12} = SI_{23} = SI_{13} = 5/91 = 0.0549 \\ SI_{123} = 1/91 = 0.0110 \end{gathered}$$

In practical applications, the actual form of the model output is often unknown, making it challenging to determine the precise order of PCE expansion. Therefore, various expansion orders need to be explored. Here, we consider PCE expansion orders of 3rd, 4th, 5th, and 6th orders. The matching points are selected using IPCM, and the resulting sensitivity indices are presented in Table 3.9.

Table 3.9 illustrates that, in comparison with the analytical solution, when the expansion order is 3, the sensitivity indices of the PCE method demonstrate good accuracy. The errors of the indices for orders 1 and 2 are only 0.2% and 1.2%, respectively, with only the error for the index of order 3 being larger. With an expansion order of 4, the errors for indices of orders 1, 2, and 3 are all reduced to 1% or less. For an expansion order of 5, the indices for order 1 are virtually identical to the exact values, while the errors for orders 2 and 3 are reduced to 0.1%. Lastly, when

Table 3.9 Polynomial function s–sensitivity indexes for different expansion orders

Sensitivity index	Analytical solution	Expansion order							
		$d = 3$		$d = 4$		$d = 5$		$d = 6$	
		Value	Error/%	Value	Error/%	Value	Error/%	Value	Error/%
SI_1	0.2747	0.2752	0.2	0.2741	0.2	0.2747	0	0.2747	0
SI_2	0.2747	0.2752	0.2	0.2741	0.2	0.2747	0	0.2747	0
SI_3	0.2747	0.2752	0.2	0.2741	0.2	0.2747	0	0.2747	0
SI_{12}	0.0549	0.0556	1.2	0.0552	0.5	0.0550	0.1	0.0549	0
SI_{23}	0.0549	0.0556	1.2	0.0552	0.5	0.0550	0.1	0.0549	0
SI_{13}	0.0549	0.0556	1.2	0.0552	0.5	0.0550	0.1	0.0549	0
SI_{123}	0.0110	0.0091	17.6	0.0109	1.0	0.0110	0.1	0.0110	0

the expansion order corresponds to the actual order of the function (6th order), the required index matches the exact value precisely, with zero error.

2. **Ishigami function**

Consider the non-linear, non-monotonic Ishigami function, for which the analytical solution of Sobol' index is:

$$\begin{aligned} &D = \tfrac{a^2}{8} + \tfrac{b\pi^4}{5} + \tfrac{b^2\pi^8}{18} + \tfrac{1}{2}, D_1 = \tfrac{b\pi^4}{5} + \tfrac{b^2\pi^8}{50} + \tfrac{1}{2}, D_2 = \tfrac{a^2}{8}, D_3 = 0 \\ &D_{12} = D_{23} = 0, D_{13} = \tfrac{8b^2\pi^8}{225}, D_{123} = 0 \end{aligned} \tag{3.32}$$

As the Ishigami function is nonlinear, determining the precise order is challenging. Therefore, PCE expansion orders of 3rd, 5th, 7th, and 9th are selected. The sensitivity indices obtained via IPCM are presented in Table 3.10.

Table 3.10 indicates that at lower PCE expansion orders ($d = 3, 5$), the main effect indices of X_1 and X_2 and the interaction effects between X_2 and X_3 may lack precision. These indices gradually converge as the expansion order $d \geq 7$, with errors across all sensitivity indices being less than 0.2%. Notably, for indices with an analytical solution of 0, the computations tend to be more accurate across different expansion orders.

According to Eq. (3.26), the sensitivity index can be decomposed based on the input variables (X_1, X_2, X_3). Taking d=9 as an illustration, the MSI SI_1 for X_1 can be dissected into the respective contributions of each $X_i^q (q = 1, 2, \cdots, 9)$, resulting in {0.1907, 0, 0.1202, 0, 0.0028, 0, 0, 0, 0}. Notably, upon observing the non-zero terms, for X_1, the model response appears to be an odd function with respect to X_1, with the linear and cubic terms being predominant. Similarly, the MSI SI_2 for X_2 can be decomposed into {0, 0.0250, 0, 0.2763, 0, 0.1339, 0, 0.008, 0}. Here, the model response is associated with X_2 as an even function, wherein the quadratic and cubic terms dominate the interaction.

In conclusion, the sensitivity analysis method based on PCE enables the swift calculation of the influence of input variables on the corresponding sensitivity index

Table 3.10 Sensitivity index of ishigami function at different expansion orders

Index	Analytical solution	Expansion order							
		$d = 3$		$d = 5$		$d = 7$		$d = 9$	
		Value	ERR/%	Value	ERR/%	Value	ERR/%	Value	ERR/%
SI_1	0.3138	0.7659	144.1	0.4055	29.2	0.3139	0.0	0.3138	0.00
SI_2	0.4424	0.0872	80.3	0.3208	27.49	0.4433	0.20	0.4432	0.18
SI_3	0	0		0		0		0	
SI_{12}	0	0		0		0		0	
SI_{23}	0.2436	0.1469	39.70	0.2737	12.36	0.2428	0.33	0.2433	0.12
SI_{13}	0	0		0		0		0	
SI_{123}	0	0		0		0		0	
TSI_1	0.5574	0.9128	63.76	0.6792	21.85	0.5567	0.13	0.5571	0.05
TSI_2	0.4424	0.0872	80.29	0.3208	27.49	0.4433	0.20	0.4432	0.18
TSI_3	0.2436	0.1469	39.70	0.2737	12.36	0.2428	0.33	0.2433	0.12

across various expansion orders. Consequently, it facilitates the determination of the predominant order of input variables.

3. **Sobol's function**

Consider Sobol' function, represented as:

$$S = \prod_{i=1}^{q} \frac{|4X_i - 2| + a_i}{1 + a_i} \tag{3.33}$$

where the input variables $X_i (i = 1, 2, \cdots, q)$ follow a uniform distribution on [0, 1].

The analytical solution of Sobol' index is [20]:

$$D = \prod_{i=1}^{q} (D_i + 1) - 1, D_i = \frac{1}{3(1 + a_i)^2}, SI_{i_1 i_2 \cdots i_s} = \frac{1}{D} \prod_{i=1}^{S} D_i \tag{3.34}$$

We select $a = [1, 2, 5, 10, 20, 50, 100, 500]$ and $q = 8$ for the Sobol' function, resulting in an eighth-order function. If subjected to an eight-variable, 8th order PCE, it would require computing 12,870 terms, which is prohibitively expensive. Thus, we initially opt for a 2nd order PCE expansion, comprising 45 terms. Due to the numerous variables, we focus solely on *MSI* and *TSI*. The Sobol' function's sensitivity index at $d = 2$ is presented in Table 3.11 and Fig. 3.8.

It is evident that the first four variables exert the most significant influence on the output response, whether analysed in terms of main effects indices or total effects indices. Consequently, the last four variables are held constant at their mean values ($X_5 = X_6 = X_7 = X_8 = 0.5$), while the first four variables maintain a uniform distribution on [0, 1], resulting in a simplified model. For this simplified model,

Table 3.11 Sensitivity index of Sobol' function at d = 2

Sensitivity index	Spiritual analysis	PCE ($d = 2$)
SI_1	0.6037	0.6263
SI_2	0.2683	0.2814
SI_3	0.0671	0.0711
SI_4	0.0200	0.0213
SI_5	0.0055	0.0062
SI_6	0.0009	0.0012
SI_7	0.0002	0.0006
SI_8	0.0000	0.0002
TSI_1	0.6342	0.6538
TSI_2	0.2945	0.3128
TSI_3	0.0756	0.0792
TSI_4	0.0227	0.0231
TSI_5	0.0062	0.0068
TSI_6	0.0011	0.0015
TSI_7	0.0003	0.0007
TSI_8	0.0000	0.0002

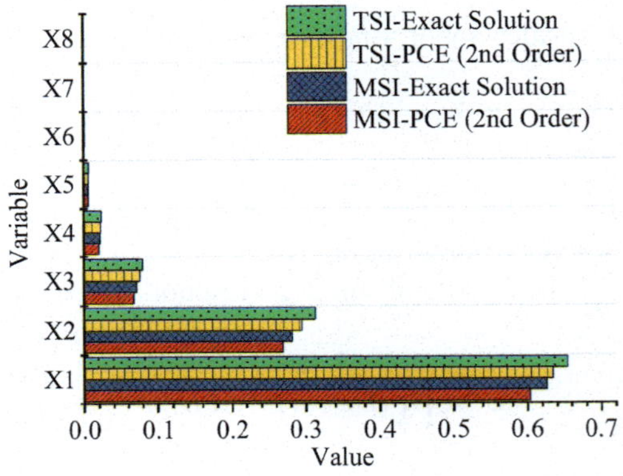

Fig. 3.8 Sensitivity index of Sobol' function at d = 2

the PCE expansion orders are set at 3rd, 5th, and 7th orders, with the matching points selected by IPCM. The sensitivity indices are detailed in Table 3.12, with the expansion terms P being 35, 126, and 330, respectively.

Table 3.12 reveals that, in comparison with the analytical solution, when $d = 3$, the computed MSI and TSI display a one-bit accuracy, albeit with relatively large relative errors. However, as $d \geq 5$, the sensitivity index progressively converges towards the exact value, with the relative errors of MSI and TSI being less than 3% at $d = 5$, and less than 1% at $d = 7$. These findings indicate that the simplified model

Table 3.12 Sensitivity index under different expansion orders for 4 parameters

Index	Analytical solution	Expansion order					
		$d = 3$		$d = 5$		$d = 7$	
		Value	Error/%	Value	Error/%	Value	Error/%
SI_1	0.6037	0.6334	4.9	0.6144	1.8	0.6085	0.8
SI_2	0.2683	0.2778	3.5	0.2680	0.1	0.2683	0
SI_3	0.0671	0.0685	2.1	0.0657	2.1	0.0666	0.7
SI_4	0.0200	0.0203	1.5	0.0194	3.0	0.0198	1.0
TSI_1	0.6342	0.6334	1.3	0.6434	1.5	0.6412	1.0
TSI_2	0.2945	0.2778	5.7	0.2931	0.5	0.2967	0.7
TSI_3	0.0756	0.0685	9.4	0.0739	2.2	0.0756	0
TSI_4	0.0227	0.0203	10.6	0.0221	2.6	0.0225	0.9

can effectively substitute the original function for sensitivity analysis while ensuring calculation accuracy.

For the two cases $d = 5$ and $d = 7$, the computational costs amount to 171 times (45 + 126) and 375 times (45 + 330), respectively. This represents a substantial reduction from the original 8-dimensional 8th-order function, which required 12,870 computations. Hence, the method demonstrates a noteworthy capability to markedly diminish the computational expense of sensitivity analysis while preserving accuracy.

This calculation example illustrates the advantages of the PCE-based sensitivity analysis method in optimal design. It showcases the method's capability to reduce dimensionality and shorten the time required for optimal design by identifying and fixing uncertainty quantities that have minimal influence on the index.

3.4 Constraint Failure Probability Solution Based on PCE Method

Solving for the failure probability of a constraint is the focal point of reliability-based optimal design. The most straightforward approach involves calculating the failure probability by integrating the PDF of the constraints within the failure domain. However, obtaining the PDF of the constraints directly poses challenges. In this section, the PCE method is combined with the MEM to directly solve the integral of the PDF of the constraints within the failure domain, thereby facilitating the solution of the failure probability.

3.4.1 Fundamentals and Expansion

The MEM [21] facilitates the determination of the PDF of a function by utilising its moments of various orders (mean, variance, skewness, kurtosis, and other higher-order moments). These moments can be derived from the polynomials of the function established through the PCE method, eliminating the need for selecting additional sample points, as required in commonly used statistical analysis methods. In the realm of reliability design, Zhang et al. [22] introduced a partial moment-based MEM to derive the PDF of structural response. Additionally, Dai et al. [23] merged the MEM with Bayesian analysis for application in uncertainty analysis of reliability. Ovalle et al. [24], on the other hand, employed the MEM for estimating the probability distribution parameter of the local load in electric vehicles.

The Shannon entropy [21] is defined as:

$$H = -c\int_{-\infty}^{+\infty} f_x(x)\ln f_x(x)\mathrm{d}x \tag{3.35}$$

where $c \geq 0$ is a constant, and $H \geq 0$.

The principle of Jaynes' maximum entropy states that given known moment constraints, there exists a distribution among all possible probability distributions that maximises the information entropy. The probability distribution obtained through this principle is minimally biased and can provide a better fit for the PDF of the functional function.

According to the principle of maximum entropy, considering the random variable x with the first m orders of origin moments $V_{x_i}(i = 1, 2, \cdots, m)$ constraints to maximise Eq. (3.35), we obtain:

$$V_{x_i} = E(x^i) = \int_{-\infty}^{+\infty} x^i f_x(x)\mathrm{d}x, i = 0, 1, 2, \cdots, m \tag{3.36}$$

Using the Lagrange multiplier method, we get:

$$L = -c\int_{-\infty}^{+\infty} f_x(x)\ln f_x(x)\mathrm{d}x + \sum_{i=0}^{m}\lambda_i[\int_{-\infty}^{+\infty} x^i f_x(x)\mathrm{d}x - V_{x_i}] \tag{3.37}$$

where $\boldsymbol{\lambda} = [\lambda_0, \lambda_1, \cdots, \lambda_m]$ is an undetermined constant.

At the extreme point, when $\frac{\partial L}{\partial f_x(x)} = 0$ A $= 0$:

$$-c\int_{-\infty}^{+\infty} [\ln f_x(x) + 1]\mathrm{d}x + \sum_{i=0}^{m}\lambda_i(\int_{-\infty}^{+\infty} x^i\mathrm{d}x) = \int_{-\infty}^{+\infty} -c[\ln f_x(x) + 1] + \sum_{i=0}^{m}\lambda_i\int_{-\infty}^{+\infty} x^i\mathrm{d}x = 0 \tag{3.38}$$

This simplifies as:

$$\ln f_x(x) + 1 - \frac{1}{c}\sum_{i=0}^{m} \lambda_i x^i = 0 \tag{3.39}$$

Assuming that $a_0 = \frac{\lambda_0}{c} - 1, a_1 = \frac{\lambda_1}{c}, \cdots, a_m = \frac{\lambda_m}{c}$, we can get:

$$f_x(x) = \exp(\frac{1}{c}\sum_{i=0}^{m} \lambda_i x^i - 1) = \exp(\sum_{i=0}^{m} a_i x^i) \tag{3.40}$$

Substituting Eq. (3.40) into Eq. (3.36), we obtain a system of nonlinear equations about $a = [a_0, a_1, \cdots, a_m]$, which is typically solved using the first 4 orders of original moments of the random variable:

$$V_{x_i} = E(x^i) = \int_{-\infty}^{+\infty} x^i \exp(\sum_{i=0}^{m} a_i x^i)\mathrm{d}x, i = 0, 1, \cdots, 4 \tag{3.41}$$

Generally, Eq. (3.41) is transformed into an unconstrained optimisation problem:

$$\min \sum_{i=0}^{4} w_i (V_{x_i}^*(a) - V_{x_i}) \tag{3.42}$$

where w_i is the weight coefficient, and $V_{x_i}^*(a)$ is the computed original moments.

Consequently, $a = [a_0, a_1, \cdots, a_m]$ can be solved, and the PDF of the random variable x can be obtained from Eq. (3.40).

It is evident that using MEM requires knowledge of the first four orders of the original moments of the response variable. For the response variable Z, given the orthogonal nature of polynomials, obtaining its first 4 orders of central moments is simple. To prevent computational overflow, Z is generally standardised to $Y = (Z - \mu_Z)/\sigma_Z$, and the central moments of Z are related to the original moments of Y, as follows:

$$E[(Z - \mu_Z)^i] = E[(\sigma_Z Y)^i] = \sigma_Z^i E(Y^i) = \sigma_Z^i V_{Y_i}, i = 0, 1, \cdots, 4 \tag{3.43}$$

$$V_{Y_1} = \frac{E(Z - \mu_Z)}{\sigma_Z} = \frac{E(Z) - \mu_Z}{\sigma_Z} = 0 \tag{3.44}$$

$$V_{Y_2} = \frac{E[(Z - \mu_Z)^2]}{\sigma_Z^2} = \frac{\sigma_Z^2}{\sigma_Z^2} = 1 \tag{3.45}$$

$$V_{Y_3} = \frac{E[(Z - \mu_Z)^3]}{\sigma_Z^3} = \frac{1}{\sigma_Z^3}\sum_{i=1}^{P-1}\sum_{j=1}^{P-1}\sum_{k=1}^{P-1} E(\Psi_i \Psi_j \Psi_k)\alpha_i \alpha_j \alpha_k \tag{3.46}$$

$$V_{Y_4} = \frac{E[(Z-\mu_Z)^4]}{\sigma_Z^4} = \frac{1}{\sigma_Z^4}\sum_{i=1}^{P-1}\sum_{j=1}^{P-1}\sum_{k=1}^{P-1}\sum_{l=1}^{P-1} E(\Psi_i\Psi_j\Psi_k\Psi_l)\alpha_i\alpha_j\alpha_k\alpha_l \tag{3.47}$$

The expectation for the product of multiple orthogonal multidimensional polynomials can be derived from Eq. (3.6) and the mutual independence of input variables:

$$\begin{aligned} E(\Psi_i\Psi_j\Psi_k) &= E\left(\prod_{t=1}^{n}\varphi_{i_t}\prod_{t=1}^{n}\varphi_{j_t}\prod_{t=1}^{n}\varphi_{k_t}\right) = E\left[\left(\varphi_{i_1}\varphi_{j_1}\varphi_{k_1}\right)\left(\varphi_{i_2}\varphi_{j_2}\varphi_{k_2}\right)\cdots\left(\varphi_{i_n}\varphi_{j_n}\varphi_{k_n}\right)\right] \\ &= \prod_{t=1}^{n} E\left(\varphi_{i_t}\varphi_{j_t}\varphi_{k_t}\right) \end{aligned} \tag{3.48}$$

$$\begin{aligned} E\left(\Psi_i\Psi_j\Psi_k\Psi_l\right) &= E\left(\prod_{t=1}^{n}\varphi_{i_t}\prod_{t=1}^{n}\varphi_{j_t}\prod_{t=1}^{n}\varphi_{k_t}\prod_{t=1}^{n}\varphi_{l_t}\right) \\ &= E\left[\left(\varphi_{i_1}\varphi_{j_1}\varphi_{k_1}\varphi_{l_1}\right)\left(\varphi_{i_2}\varphi_{j_2}\varphi_{k_2}\varphi_{l_2}\right)\cdots\left(\varphi_{i_n}\varphi_{j_n}\varphi_{k_n}\varphi_{l_n}\right)\right] \\ &= \prod_{t=1}^{n} E\left(\varphi_{i_t}\varphi_{j_t}\varphi_{k_t}\varphi_{l_t}\right) \end{aligned} \tag{3.49}$$

Using the Hermite polynomials as an illustration, the expectation of the product of orthogonal polynomials with identical input variables can be expanded according to Eq. (3.50), as follows:

$$\begin{aligned} E\left(\varphi_{i_t}\varphi_{j_t}\cdots\varphi_{l_t}\right) &= E\left(a_1\varepsilon_t^1 + a_2\varepsilon_t^2 + \cdots + a_m\varepsilon_t^m\right) \\ &= a_1E\left(\varepsilon_t^1\right) + a_2E\left(\varepsilon_t^2\right) + \cdots + a_mE\left(\varepsilon_t^m\right) \end{aligned} \tag{3.50}$$

$$E(\varepsilon^b) = \begin{cases} 0, & b \text{ is an odd number} \\ (b-1)!!, & b \text{ is an even number} \end{cases} \tag{3.51}$$

Therefore, when the response variable Z is expanded using the PCE method, the first 4th order moments of origin of the normalised Y can be determined.

In summary, regarding the response variable Z, the coefficients and bases of its PCE encapsulate the first four orders of the moments of origin. Subsequently, the MEM is employed to resolve the PDF of Z. The probability of failure can then be derived by integrating the PDF:

$$p_f = \Pr(Z \le 0) = \Pr\left(Y \le -\frac{\mu_Z}{\sigma_Z}\right) = \int_{-\infty}^{-\frac{\mu_Z}{\sigma_Z}} \exp\left(\sum_{i=0}^{4} a_iY^i\right)\mathrm{d}Y \tag{3.52}$$

3.4.2 Numerical Examples

1. **Using MEM to solve for randomly distributed PDFs**

To verify the accuracy of the probability distribution function fitted by MEM, three typical random distributions are selected for fitting in this book.

(1) Standard normal distribution, i.e.

$$f(x) = \frac{1}{\sqrt{2\pi}} e^{-\frac{x^2}{2}} \quad , \quad -\infty < x < +\infty \tag{3.53}$$

(2) Exponential distribution, i.e.

$$f(x) = \begin{cases} \frac{1}{\theta} e^{-\frac{x}{\theta}}, & x > 0 \\ 0, & x \leq 0 \end{cases} \tag{3.54}$$

where θ=1 is chosen.

(3) Greatest value distribution, i.e.

$$f(x) = \frac{1}{\sigma} e^{-\frac{x-\mu}{\sigma}} \exp\left(-e^{-\frac{x-\mu}{\sigma}}\right) \tag{3.55}$$

Here $\mu = 0$ and $\sigma = 1$ are selected.

The first 4 order moments of origin for the three typical distributions are shown in Table 3.13.

The PDF coefficients obtained through MEM are shown in Table 3.14.

Table 3.13 First 4 origin moments of three typical distributions

Type of distribution	Origin moments			
	1st order	2nd order	3rd order	4th order
Standard normal distribution	0	1	0	3
Exponential distribution	1	2	6	24
Maximum distribution	0.5772	1.9781	5.4449	23.5615

Table 3.14 The PDF coefficients obtained by MEM

Type of distribution	Coefficients				
	a_0	a_1	a_2	a_3	a_4
Standard normal distribution	−0.9189	0	−0.5000	0	0
Exponential distribution	−1.1934	2.3411	−2.3504	0.5631	−0.0424
Maximum distribution	−0.9893	0.1348	−0.5650	0.1157	−0.0078

For the standard normal distribution, the PDF obtained from the fitting process is:

$$f_{MEM}(x) = e^{-0.9189-0.5x^2} = e^{-0.9189}e^{-\frac{x^2}{2}} = \frac{1}{\sqrt{2\pi}}e^{-\frac{x^2}{2}} \tag{3.56}$$

It is identical to the probability function of the standard normal distribution.

For the exponential and extreme value distributions, assessing the accuracy of the fitting based solely on the MEM-derived coefficients may be insufficient. Therefore, this book employs visual comparisons by plotting PDF curves and CDF curves. This approach allows for a more intuitive assessment of the differences between the distributions calculated by MEM and the standard distribution. These visual comparisons are illustrated in Figs. 3.9 and 3.10.

In the case of the exponential distribution, the PDF is inherently a segmented function. Consequently, the probability density curves fitted by MEM may exhibit inaccuracies, particularly in proximity to $x = 0$, resulting in significant errors. However,

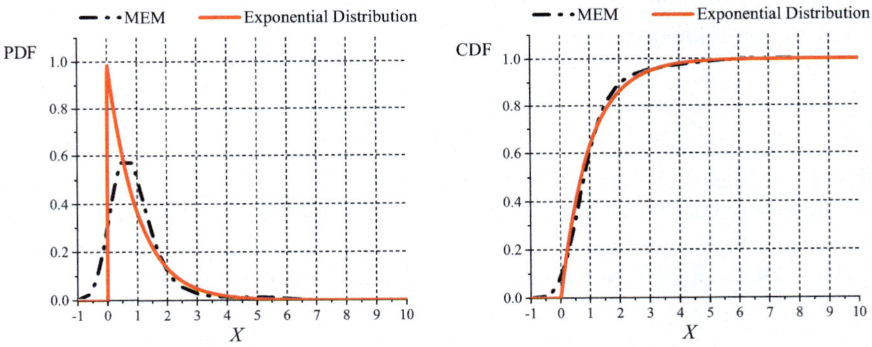

Fig. 3.9 Comparison of exponential distribution PDF and CDF curves

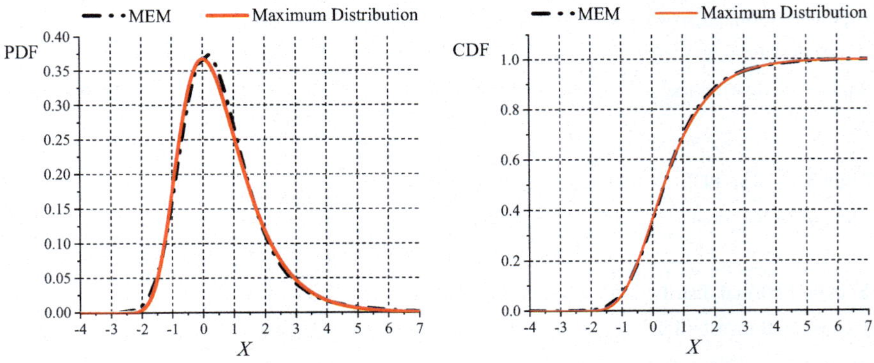

Fig. 3.10 Comparison of the PDF and CDF curves of the extreme value distribution

the cumulative density curves tend to be more accurate. Notably, discrepancies with the standard distribution primarily occur within the range of $x < 0$.

In the case of the extreme value distribution, the probability density and cumulative density curves fitted by MEM demonstrate a high level of accuracy and alignment with the expected values. Notably, the cumulative density curve exhibits a remarkable degree of overlap with the standard distribution curve, indicating a robust fit.

In summary, for the aforementioned three commonly used distributions, the MEM fits generally exhibit superior accuracy, with the exception of the exponential distribution, where the PDF curve does not achieve a satisfactory level of fit.

2. **Combination of PCE method and MEM to solve the failure probability of the response**

The 2-dimensional polynomial function of order 3, as discussed in Sect. 3.2.2, serves as an illustration to assess the precision of the PCE method in forecasting the first four orders of moments of the response, as discussed in Sect. 3.4.2. The outcomes obtained from the MC method are considered as the benchmark, while the PCE expansion is conducted at the third order. The findings are presented in Table 3.15.

Table 3.15 reveals that in comparison with the outcomes derived from the MC method, the discrepancies in the first four orders of moments of the response remain under 1%. The first four orders of origin moments following the normalisation of the response are depicted in Table 3.16.

The PDF of the function is fitted by the MEM, with the coefficients presented in Table 3.17. The PDF curve of the function 'y' after normalisation is depicted in Fig. 3.11.

Table 3.15 First 4-step moments obtained by PCE and MC methods

	Mean value	Standard deviation	Coefficient of Skew	Coefficient of Kurtosis
3rd order PCE	4682.30	4227.46	1.8803	8.6362
MC	4684.59	4231.60	1.8872	8.6841
Error/%	0.06	0.09	0.66	0.94

Table 3.16 First 4th order moments of origin after response normalisation

	1st order	2nd order	3rd order	4th order
3rd order PCE	0	1	1.8803	8.6362

Table 3.17 PDF coefficients obtained by MEM

a_0	a_1	a_2	a_3	a_4
−0.7082	−0.7611	−0.8524	0.3429	−0.0356

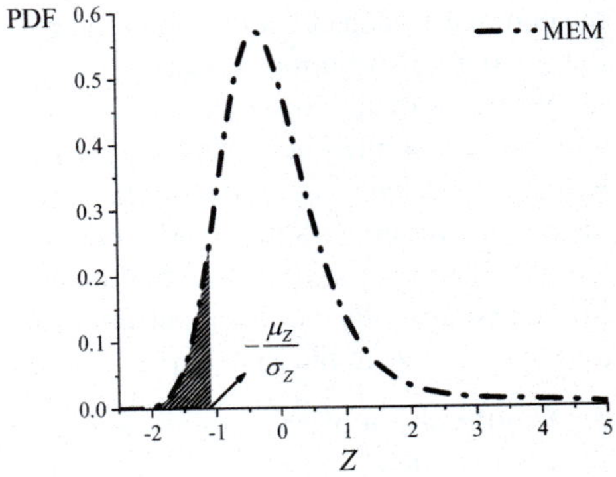

Fig. 3.11 PDF curve after function standardisation

Table 3.18 Comparison of probability of failure by PCE and MC methods

	MEM	MC	Error/%
Failure probability p_f	0.0647	0.0636	1.7

The probability of failure (shaded area) can be obtained by integrating the aforementioned PDFs. A comparison of the probability of failure obtained by the PCE method and the MC method is presented in Table 3.18.

The error in the failure probability calculated by MEM compared to the MC method is 1.7%, meeting the requirements for engineering applications.

3.5 Application of the Improved PCE Method to the Ship Arithmetic Example

3.5.1 Example of an Optimised Design for a Bulk Carrier

1. **Description of the optimizing problem**

In this section, the previously discussed methods are applied to an example concerning the uncertainty optimisation design of a ship. Specifically, we consider the bulk carrier engineering optimisation model proposed by Hannapel et al. The model undergoes optimisation under deterministic and uncertainty conditions, respectively. The structure of the bulk carrier optimisation model is presented in Table 3.19, with the conversion factor of 1kn = 1.852 km/h. Detailed formulae relevant to the specific model are provided in Appendix B.

2. **Uncertainty optimisation design process for bulk carriers**

Figure 3.12 shows the uncertainty optimisation process, which comprises the following steps:

Table 3.19 Optimisation model for bulk carriers

Design variable	
Hull length *L*/m	$150 \le L \le 274.32$
Hull breadth *B*/m	$20 \le B \le 32.31$
Hull depth *D*/m	$13 \le D \le 25$
Hull draft *T*/m	$10 \le T \le 11.71$
Block coefficient C_B	$0.63 \le C_B \le 0.75$
Design speed V_k/kn	$14 \le V_k \le 20$
Constraint condition	
$L/B \ge 6$	
$L/D \le 15$	
$L/T \le 19$	
$F_n \le 0.32$	
$25000 \le DW \le 500000$	
$T - 0.45DW^{0.31} \le 0$	
$T - 0.7D - 0.7 \le 0$	
$0.07B - 0.35T - \frac{(0.085C_B - 0.002)B^2}{TC_B} + 1 + 0.25D \le 0$	
Objective function	
Costs of freight *TC*/(£/yr)	$TC = C_A/AC$

(1) Preparation work. Determine the optimal expansion order of the original model PCE, conduct sensitivity analysis on uncertain variables, and rank them based on their influence. Fix uncertain variables with lesser impact to reduce model dimensionality. Subsequently, determine the optimal expansion order of PCE for the reduced model.
(2) Selection of design variables: Determine the range of values for each design variable.
(3) Selection of design variables and defining their values: Select a set of design variables (bulk carrier scheme) and input them into the inner uncertainty analysis module.
(4) Uncertainty point selection: Utilise the optimal expansion order of PCE, along with the dimension and distribution of uncertainty, to employ the IPCM method for selecting uncertainty points.
(5) PCE expansion: Conduct PCE expansion on the objective and constraint functions.
(6) Uncertainty optimisation: Based on design requirements, select the type of uncertainty optimisation needed and obtain the mean and standard deviation of the objective function, as well as the failure probability of the constraint. Pass these parameters to the optimiser of the outer layer of optimisation.
(7) Iterative process: The optimiser determines if the entire optimisation process meets the preset number of iterations. If the criteria are met, the uncertainty

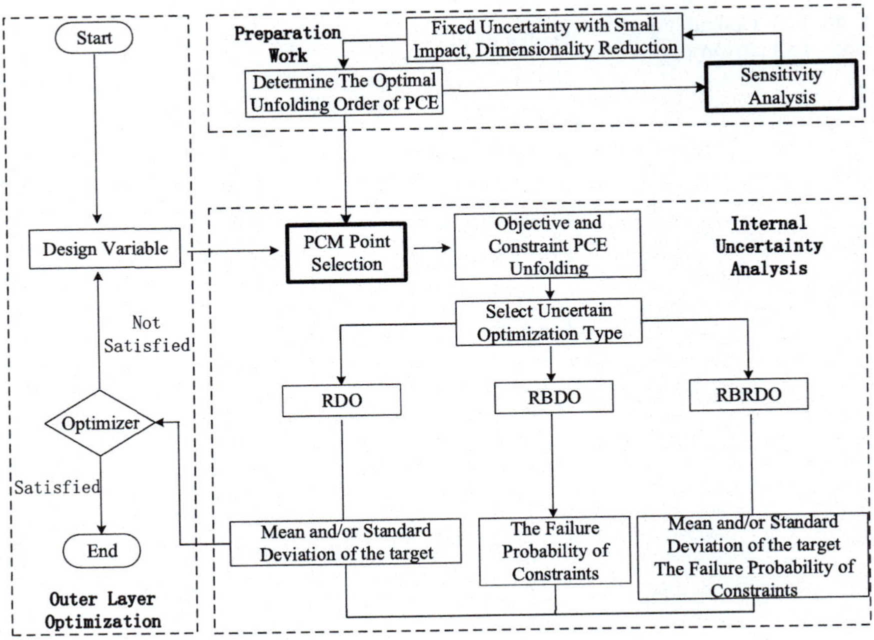

Fig. 3.12 The process of uncertainty optimisation

optimisation process concludes. If not, select a new set of design variables and repeat steps (3) to (6).

3. **Optimisation problem description**

(1) Deterministic optimisation.

For ease of comparison with the results of uncertainty optimisation, this example initially conducts deterministic optimisation utilising a multi-island genetic algorithm with a subgroup size of 10, 10 islands, and a genetic iteration count of 50. Design variables, constraints, and objective functions are selected as outlined in Table 3.19. The optimisation outcomes alongside the initial plan are presented in Table 3.24.

(2) Uncertainty optimisation.

(1) The selection of uncertain quantities.

During sea navigation, the ship's power remains constant. However, due to the influence of wind and waves, the ship encounters increased resistance within the waves, leading to heightened movement and a consequent reduction in actual speed, often preventing it from reaching the intended design speed. Additionally, in hazardous sea conditions, crew members may manually adjust the ship's speed. Consequently, the actual speed of a ship is not fixed but rather fluctuates around the design speed, presenting an element of uncertainty. This variability in ship speed can be expressed as:

$$V_{actual} = V_k + Z \tag{3.57}$$

where V_{actual} is the actual speed; V_k is the design speed; and Z represents the speed fluctuation, following a normal distribution characterised by a mean of 0 and a standard deviation of $0.1V_k$.

The weight of steel for bulk carriers is determined by:

$$W_S = 0.034L^{1.7}B^{0.7}D^{0.4}C_B^{0.5} \tag{3.58}$$

For the engineering optimisation model of the bulk carrier, the calculation formula for steel weight is derived from statistical regression of actual ship data. However, due to the limited sample size of the actual ship data and the unknown regression method employed, it is challenging to control errors and accurately reflect the actual engineering scenario using regression index values. Consequently, the index of the calculation formula is considered an uncertain quantity. To address this uncertainty, the uncertainties X (normal distribution with a mean of 1.7 and a standard deviation of 0.17), Y (normal distribution with a mean of 0.7 and a standard deviation of 0.07), A (normal distribution with a mean of 0.4 and a standard deviation of 0.04), and B (normal distribution with a mean of 0.5 and a standard deviation of 0.05) are defined. Through this definition, we can determine:

$$W_S = 0.034L^X B^Y D^A C_B^B \tag{3.59}$$

(2) Determination of the optimal expansion order for PCE.

In deterministic optimisation schemes, it is essential to account for the influence of uncertainty on constraints and optimisation objectives. This entails obtaining the undetermined coefficient values of the 2nd, 3rd, and 4th order PCE expansions, as depicted in Fig. 3.13.

From this analysis, it becomes evident that for $T - 0.45DW^{0.31}$, the undetermined coefficients beyond number 72 are predominantly zero, suggesting that truncating to the 4th order is sufficient to achieve accuracy. Similarly, for DW, the coefficients for the 2nd, 3rd, and 4th order expansions exhibit minimal variation, with the 2nd order expansion adequately meeting its accuracy requirements. In the case of DW, the coefficients for the 3rd order expansion mostly approach zero, and beyond the 56th coefficient (4th order terms), they are consistently close to zero, indicating that the 3rd order expansion generally fulfils accuracy criteria. Conversely, for TC, starting from coefficient number 107, the values approach zero, suggesting that the 4th order expansion largely satisfies accuracy standards. In summary, based on this example, the optimal order for PCE expansion is determined to be 4.

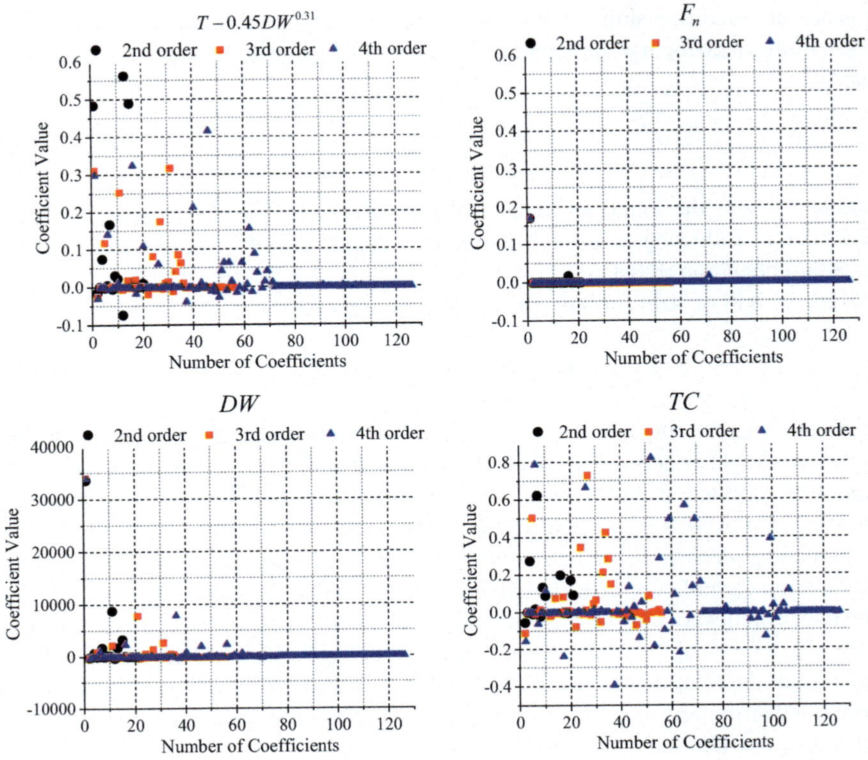

Fig. 3.13 Numerical diagram of constraints and optimisation objective coefficients affected by uncertainty

The sensitivity analysis of targets and constraints will be carried out utilising the PCE method, and the outcomes will be juxtaposed with those derived from the conventional Sobol' method. By analysing the undetermined coefficients obtained under the 4th order expansion, the sensitivity indices of the targets and constraints to X, Y, Z, A, and B can be determined. Table 3.20 and Fig. 3.14 display these results, revealing that, in comparison with the traditional Sobol' method, the sensitivity index obtained via the PCE method closely aligns with it, demonstrating accuracy that meets the stipulated requirements.

Utilising the MSI and TSI, the relative significance of the five uncertain variables can be ascertained, enabling their prioritisation. In the case $T - 0.45DW^{0.31}$, DW, and TC, X and Y emerge as predominant factors, whereas A exerts a comparatively minor influence, rendering the effects of other variables negligible. Conversely, in the case of F_n, Z predominates in influence, overshadowing the effects of other variables. This finding corroborates with F_n. Additionally, across all constraints and objectives, the MSI and TSI values for B are close to 0, rendering them negligible.

In summary, X, Y, and Z emerge as the primary influencers on both the objective and the constraints, as discerned from both the main effect index and the total effect

Table 3.20 Sensitivity index for objective and constraints

	$T - 0.45DW^{0.31}$		DW		F_n		TC	
	PCE	Sobol'	PCE	Sobol'	PCE	Sobol'	PCE	Sobol'
MSI_X	0.8692	0.8857	0.8842	0.8302	0	0	0.8507	0.8925
MSI_Y	0.0615	0.0637	0.0595	0.0582	0	0	0.0649	0.0712
MSI_Z	0.0002	0.0012	0.0002	0.0004	0.9999	0.9249	0.0023	0.0021
MSI_A	0.0136	0.0153	0.0128	0.0139	0	0	0.0148	0.151
MSI_B	0.0006	0.0005	0.0005	0.0005	0	0	0.0007	0.0006
TSI_X	0.9238	0.9159	0.9267	0.9451	0	0	0.9169	0.9237
TSI_Y	0.1060	0.1155	0.0945	0.1151	0	0	0.1180	0.1244
TSI_Z	0.0002	0.0003	0.0002	0.0002	0.9999	0.9249	0.0027	0.0031
TSI_A	0.0241	0.0230	0.0208	0.0314	0	0	0.0279	0.0293
TSI_B	0.0011	0.0020	0.0009	0.0012	0	0	0.0013	0.0021

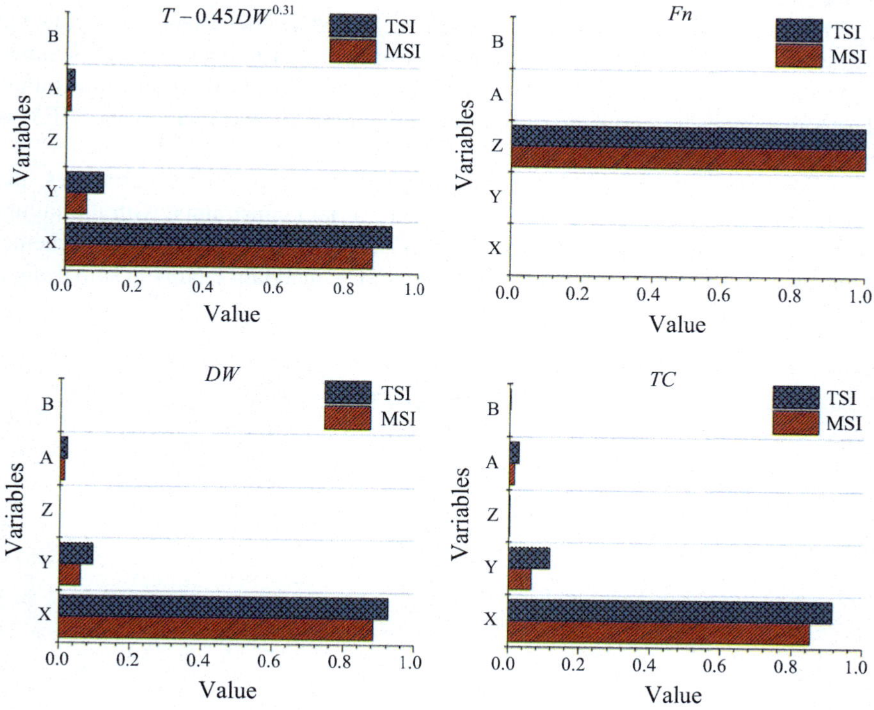

Fig. 3.14 The constraints and optimisation objectives affected by uncertainty

index analyses. Consequently, A and B are maintained at their mean values, while X, Y, and Z retain their original distributions. This simplified model serves as a substitute for the original model in the optimisation process.

For this simplified model, the procedure of determining the optimal PCE order is reiterated, revealing that the 3rd-order expansion is optimal for the model. Under the influence of three uncertainties, the 3rd-order PCE expansion (requiring 20 sample points) is utilised to determine the mean and standard deviation of the target or constrained in the deterministic optimisation optimal solution. These results are then compared with those obtained from the MC method (which necessitates 100 sample points), as depicted in Table 3.21.

Based on the findings presented in Table 3.21, with the results of the MC method regarded as the benchmark, the error in the mean and standard deviation calculated using the PCE method falls within 2 and 3.5%, respectively. This underscores the clear advantage of PCE in computational efficiency, notably evidenced by the substantial reduction in required sample points without a significant compromise in calculation accuracy.

The next step involves further processing the PCE coefficients, applying the MEM to derive the PDF of the constraints in the deterministic optimal scheme, and subsequently integrating to determine the failure probability of the constraints. The constraint failure probability p_f obtained by both the MC method and MEM is outlined in Table 3.22.

As depicted in Table 3.22, when considering the impact of uncertainties, the deterministic optimal solution may contravene certain constraints, with the failure probability of some constraints exceeding 0.1%, leading to the failure of the scheme. Consequently, it becomes imperative to account for uncertain factors in ship design.

Table 3.21 The objective and the constraint under the influence of uncertainty

The objective/the constraint		TC	$F_n - 0.32$	DW	$T - 0.45\,DW^{0.31}$
Mean	MC	10.06	0.1699	$3.408*10^4$	0.309
	PCE	10.13	0.1699	$3.479*10^4$	0.304
	Error/%	0.69	0	1.95	1.64
Standard deviation	MC	4.21	0.017	8009	0.988
	PCE	4.29	0.017	7968	1.021
	Error/%	1.90	0	0.52	3.18

Table 3.22 The failure probability p_f for MCE and MEM

Constraints	Failure probability p_f/%	
	MC	MEM
$F_n \leq 0.32$	0	0
$25000 \leq DW \leq 500000$	47.0	45.2
$T - 0.45DW^{0.31} \leq 0$	11.2	10.6

Table 3.23 Uncertainty optimisation design variables, constraints, and optimisation objectives

Item	RDO	RBDO	RBRDO
Design variables	Design variables in Table 3.19		
Uncertainty variables	$X: \mu_X = 1.7, \sigma_X = 0.1\,\mu_X$ $Y: \mu_Y = 0.7, \sigma_Y = 0.1\,\mu_Y$ $Z: \mu_Z = 0, \sigma_Z = 0.01\,V_k$		
Constraints	Table 3.19 constraints	$p_f(F_n \leq 0.32) \leq 1\%$ $p_f(25000 \leq DW \leq 500000) \leq 1\%$ $p_f(T - 0.45DW^{0.31} \leq 0) \leq 1\%$	
Objectives	$\min \mu(TC)/\sigma(TC)$	$\min TC$	$\min \mu(TC)$ and $\sigma(TC)$

Moreover, in comparison with the results derived from the MC method, the discrepancy in the failure probability obtained through the MEM is minimal. Hence, the proposed method offers a viable alternative to the MC method for computing the failure probability.

(3) Parameter setting of uncertainty design optimisation.

RDO, RBDO, and RBRDO all utilise the NSGA-II algorithm, employing a population size of 40 and 40 generations. Design variables, constraints, and objective functions are detailed in Table 3.23. Both pre- and post-dimension reduction models, comprising a 5-dimensional 4th-order (126 sample points) PCE and a 3-dimensional 3rd-order (20 sample points) PCE, are employed for UDO and validated against the MC method (100 sample points). Results of these two methods are presented in Table 3.24, and the iteration of the objective is depicted in Figs. 3.15, 3.16, 3.17, where llb = 0.453 kg.

4. Analysis of optimisation results.

From Figs. 3.15, 3.16, 3.17, it is evident that in the case of RDO, the 5D PCE model starts to converge around the 900th iteration, while the 3D PCE model begins to converge approximately at the 400th iteration. This suggests that the model, after dimension reduction, achieves convergence faster than the original model. In contrast, for RBDO and RBRDO, where solving the failure probability adds complexity, the convergence rate of the dimension-reduced model is nearly identical to that of the original model. However, considering the total time required to complete the entire optimisation process, reducing the model from 5 to 3 dimensions significantly decreases the time needed. This reduction is attributed to the decrease in collocation points from 126 to 21 in the dimension-reduced model. This reflects the notable benefits of dimension reduction. Moreover, Table 3.24 indicates minimal differences in objective values before and after dimension reduction. Specifically, the discrepancies in three cases are 1.2, 2.7, and 1.8%, respectively. This marginal variation can be attributed to the relatively small TSI of variables *A* and *B*, indicating their negligible influence. Consequently, fixing variables *A* and *B* has an extremely limited impact

Table 3.24 Comparison of optimisation results

Item	Initial	DO	RDO		
			PCE(5D)	PCE(3D)	MC(3D)
$TC/(\text{Pounds/yr})$	9.926	8.446	8.729	8.488	8.495
$\mu(TC)/(\text{Pounds/yr})$	–	10.130	9.906	9.788	9.800
$\sigma(TC)$	–	4.29	3.509	3.552	3.567
L/m	195.000	188.610	181.121	178.329	178.612
B/m	32.310	31.340	32.305	29.699	29.621
D/m	20.000	15.960	15.772	15.734	15.733
T/m	10.500	11.710	11.705	11.710	11.707
C_B	0.700	0.640	0.748	0.750	0.749
V_k/kn	16.000	14.210	14.425	14.002	14.010
$p_f(F_n \leq 0.32)$	–	0	0	0	0
$p_f(25000 \leq DW \leq 500000)$	–	10.6%	17.9%	12.1%	10.8%
$p_f(T - 0.45DW^{0.31} \leq 0)$	–	45.2%	41.2%	34.3%	34.3%

Item	RBDO			RBRDO		
	PCE(5D)	PCE(3D)	MC(3D)	PCE(5D)	PCE(3D)	MC(3D)
$TC/(\text{lb/a})$	8.811	8.576	8.583	8.843	8.603	8.639
$\mu(TC)/(\text{lb/a})$	10.128	10.077	10.091	10.227	10.042	10.019
$\sigma(TC)$	4.237	4.128	4.147	4.229	4.088	4.077
L/m	205.357	197.358	197.873	200.666	195.025	191.701
B/m	32.310	32.309	32.253	31.420	32.299	31.786
D/m	15.719	14.521	14.529	15.784	14.518	14.316
T/m	10.769	10.859	10.854	11.460	10.831	10.704
C_B	0.747	0.749	0.750	0.745	0.749	0.750
V_k/kn	14.205	14.004	14.007	14.416	14.083	14.021
$p_f(F_n \leq 0.32)$	0	0	0	0	0	0
$p_f(25000 \leq DW \leq 500000)$	0	0	0	0.6%	0.3%	0.1%
$p_f(T - 0.45DW^{0.31} \leq 0)$	0	0	0	0.5%	0.4%	0.1%

on the optimal solution's quality. In conclusion, the dimension-reduced model can effectively substitute the original model for uncertainty design optimisation.

Table 3.24 provides the optimal solutions after dimension reduction of RDO, RBDO, and RBRDO, thereby affirming the feasibility of PCE. Additionally, the advantage of the PCE method is evident from Figs. 3.15, 3.16, 3.17. PCE demonstrates the capability to maintain accuracy with fewer sample points. In the case of RDO, the objective function solved by both the PCE and MC methods. The optimal solutions obtained by these two methods exhibit remarkable similarity, of PCE exhibits a significant reduction early in the iteration process, converging around the 400th iteration. Specifically, the objective function of PCE completes the entire

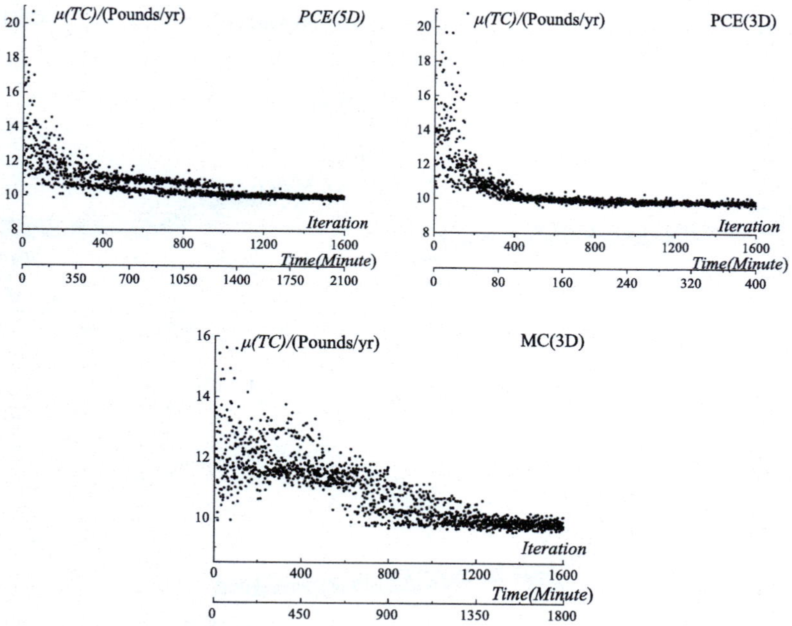

Fig. 3.15 Comparison of objective values in RDO process

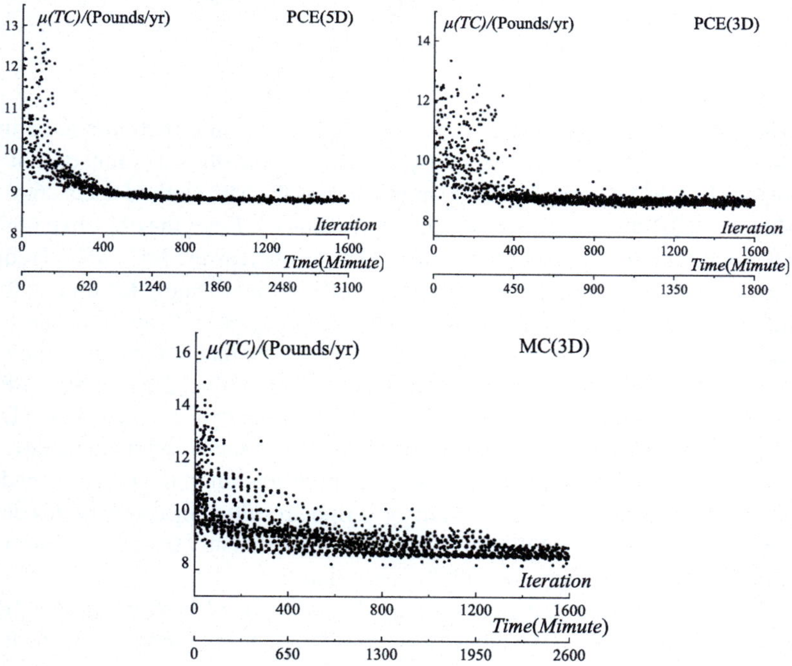

Fig. 3.16 Comparison of objective values in RBDO process

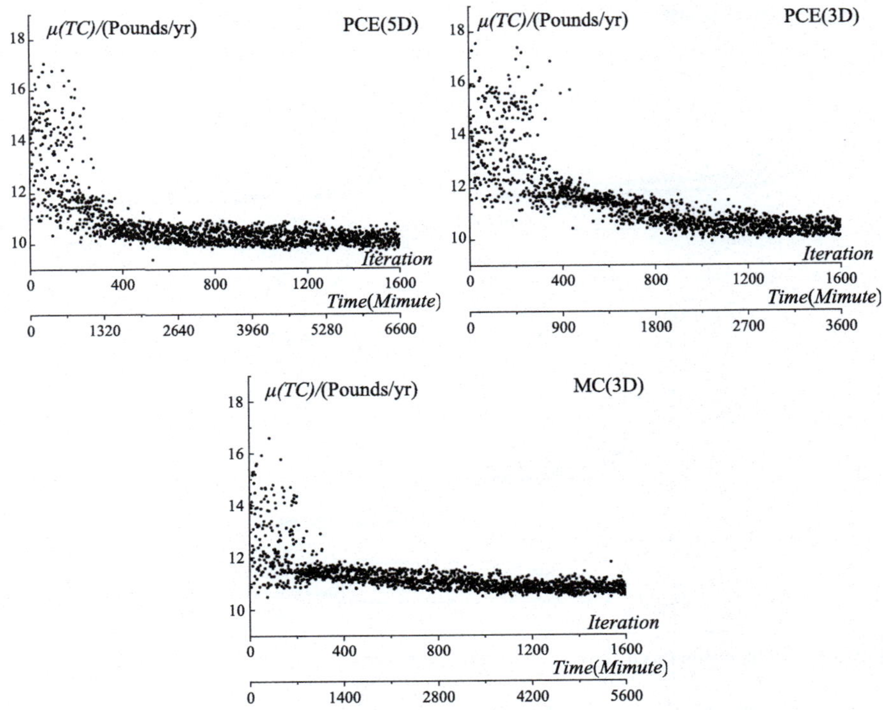

Fig. 3.17 Comparison of objective values in RBRDO process

1600 iterations in 400 min, whereas the MC method requires 1800 min to complete the same number of iterations. Similarly, for RBDO, the objective function of PCE completes the 1600 iterations in 1800 min, whereas the MC method takes 2600 min. Notably, for RBRDO, the disparity is more pronounced. Here, the objective function of PCE completes the 1600 iterations in 3600 min, whereas the MC method requires 5600 min. Overall, the computational time of PCE is significantly lower than that of the MC method, demonstrating clear advantages in engineering applications.

Using the result of PCE as a reference, it is observed that compared to the initial design, the TC in the three uncertain optimal cases is reduced by 14.5% in RDO, 13.6% in RBDO, and 13.2% in RBRDO, which is marginally lower than that of DO, at 14.9%. Notably, in RDO, the optimal case exhibits lower mean and standard deviation values compared to RBDO, indicating that the perturbation of the objective induced by uncertainty is considerably smaller in the former, resulting in a more robust design. In RBDO, the failure probability of constraints is consistently 0, which is lower than that of RDO, implying a design with higher reliability.

In the case of RBRDO, it amalgamates the strengths of both RDO and RBDO. Here, both the mean and the standard deviation experience reductions, and the failure probability of constraints is below 0.5%. This underscores the capability of the optimal scenario in RBRDO to meet reliability requirements while simultaneously mitigating the perturbation induced by uncertainty. Consequently, it ensures a balance between reliability and robustness in the design.

3.5.2 KCS Profile Optimisation of Standard Hull Form

1. KCS hull form description

In this study, the KRISO Container Ship (KCS), a standard ship hull form published by ITTC, has been chosen as the focal point of investigation. The geometric representation of the KCS is depicted in Fig. 3.18. Objective functions are established based on the mean and standard deviation of the EEDI of the ship, with the IMO stipulated requirements for EEDI serving as constraints. Moreover, the study evaluates the ship's performance under the influence of uncertainty. Detailed principal ship form parameters for the KCS model are provided in Table 3.25.

2. The EEDI formula

According to Resolution MEPC. 245(66) [25], EEDI is expressed as:

$$EEDI = \frac{E_{ME} + E_{AE} + E_{PT} + E_{eff}}{f_i \cdot capacity \cdot V_{ser} \cdot f_w} \tag{3.60}$$

$$\begin{aligned}
E_{ME} &= \left(\prod_{j=1}^{n} f_j\right)\left(\sum_{i=1}^{nME} P_{ME(i)} \cdot C_{FME(i)} \cdot SFC_{ME(i)}\right) \\
E_{AE} &= \sum_{i=1}^{nME} P_{AE} \cdot C_{FAE} \cdot SFC_{AE} \\
E_{PT} &= \left(\prod_{j=1}^{n} f_j \sum_{i=1}^{nPTI} P_{PTI(i)} - \sum_{i=1}^{neff} f_{eff\,(i)} P_{AEeff\,(i)} \cdot C_{FME} \cdot SFC_{ME}\right) \\
E_{eff} &= -\left(\sum_{i=1}^{neff} f_{eff\,(i)} P_{AEeff\,(i)} \cdot C_{FME} \cdot SFC_{ME}\right)
\end{aligned} \tag{3.61}$$

Fig. 3.18 KCS diagram

Table 3.25 Principal ship form parameters of the KCS

Item	L_{pp}/m	B_{wl}/m	D/m	T/m	C_b	F_r
Model	7.2786	1.019	0.6013	0.3418	0.651	0.26
Full scale ship	230.0	32.2	19.0	10.8	0.651	0.26

Table 3.26 Relevant parameters in EEDI formula

Parameter	Value	Parameters	Value
P_{ME}	$75\%MCR_{ME}$	SFC_{ME}/SFC_{AE}	200
P_{AE}	$5\%MCR_{ME}$	$f_i/f_j/fw/f_{eff}$	1
C_F	3.1144	$P_{AEeff}/P_{PTI}/P_{eff}$	0

where, E_{ME}, E_{AE}, E_{PT}, and E_{eff} are the CO_2 emissions of the main engine, auxiliary engine, shaft device, and energy-saving equipment, respectively; SFC is the fuel consumption at 75% rated power (g/($KW.h$)); P is power for the main or auxiliary engine (KW); Capacity is the deadweight tons (t); f_j is the modifying factor for ship special design, f_i is the modifying factor for ice strengthened ships; f_{eff} is the innovation factor; and f_w is the wind wave correction factor.

The specific parameters used in the EEDI formula for the KCS are detailed as Table 3.26.

$\mathrm{MCR_{ME}}$ is the rated power of the main engine. By substituting the parameters from Table 3.26 into Eq. (3.60), (3.61), the simplified EEDI formula is:

$$
\begin{aligned}
EEDI &= \frac{E_{ME}+E_{AE}}{capacity \times V_{ser}} = C_F \times \frac{P_{ME}\times SFC_{ME}+P_{AE}\times SFC_{AE}}{capacity\times V_{ser}} \\
&= C_F \times \frac{SFC(75\%MCR_{ME}+5\%MCR_{ME})}{capacity\times V_{ser}} = C_F \times \frac{0.8\times SFC\times R_T}{capacity\times PC\times 1.94\times 10^3}
\end{aligned}
\tag{3.62}
$$

where R_T is the total resistance of the KCS, and N, $PC = 0.74$ is the propulsion coefficient according to the experimental data of the KCS.

The deadweight, can be calculated as:

$$
capacity = \Delta - W_t = \rho\nabla - W_h - W_o - W_m \tag{3.63}
$$

where $W_{t,}$ is light weight, t, which is divided into hull weight W_h, outfit weight W_o, and electromechanical weight W_m.

According to the routine statistical data of this ship type [26], W_h and W_o can be estimated based on the principal dimensions related to the principal dimensions, whereas W_m can be set according to the power of the main engine.

3. Optimise the problem description.

(1) Optimise the objective

The objective of DO is to achieve optimal hull forms with the minimum EEDI value at the design speed (Fr = 0.26). Hence, the primary aim of DO is to minimise the EEDI value. In contrast, for RBRDO, the objective is not only to enhance the objective performance but also to decrease its sensitivity to uncertainty. Therefore, both the mean and standard deviation of the EEDI value are selected as the objective functions, as detailed in Table 3.27.

Table 3.27 DO versus RBRDO optimisation objectives

Item	DO	RBRDO
Objective function	min *EEDI*	min $\mu(EEDI)/\sigma(EEDI)$

(2) Design variables

The new ship hull form is generated using a hull modification method based on RBF interpolation. Six adjustable control points are designated in the ship's bow and labelled as 1–6. The positions of these control points are illustrated in Fig. 3.19. In the diagram, control point 1 governs changes in the X (longitudinal) and Z (vertical) directions, whereas control points 2 to 6 influence changes in the Y (transverse) direction. Control point 1 encompasses two variables, namely the length parameter and the height parameter of the bulbous bow. Control point 2 controls the width parameters of the bulbous bow, while control points 3 and 5 regulate the water plane inlet angle and the shape of the inlet section. Control points 4 and 6, on the other hand, manage the shape of the bow bilge. Design variables for both DO and RBRDO are detailed in Table 3.28.

Considering hull smoothness and construction feasibility, fixed points are chosen on the control points located along the centreline, deck edge line, ship bottom line, and sections near the midship.

(3) Constrain t condition

The constraints for the DO primarily involve geometric considerations, encompassing displacement ($\triangle$) and the longitudinal position of the centre of buoyancy (X_B). Table 3.29 outlines the constraints for both DO and RBRDO.

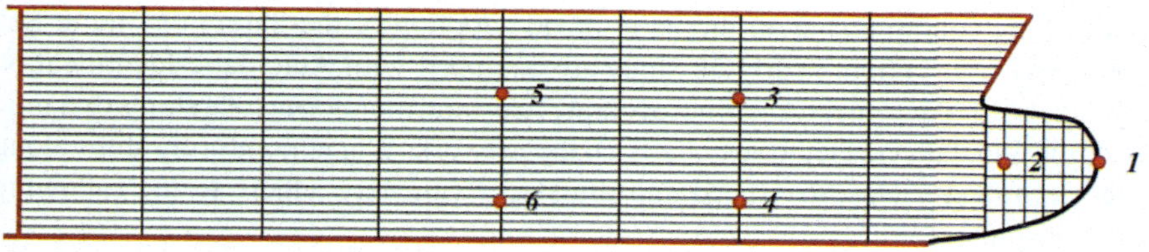

Fig. 3.19 Location of variable and fixed control points

Table 3.28 DO and RBRDO design variables

Design variable	DO	RBRDO
X_1	[−0.03, 0.02]	
Z_1	[0.175, 0.225]	
Y_2	[0.055, 0.09]	
Y_3	[0.131, 0.163]	
Y_4	[0.079, 0.129]	
Y_5	[0.431, 0.471]	
Y_6	[0.32, 0.37]	

Table 3.29 Constraints on the DO and RBRDO

Constraint condition	DO	RBRDO
∇/m^3	Maximum change ± 0.5%	
X_B/L_{pp}%	Maximum change ± 0.5%	
Reliability constraints	$P_f(EEDI < EEDI(I)) \leq 5\%$ (only RBRDO)	

For RBRDO, aside from geometric constraints, it is essential to incorporate constraints related to reliability requirements. In terms of reliability, to adhere to the energy efficiency standards set by the IMO, the ship is classified into three levels, namely GreenShip I, GreenShip II, and GreenShip III, with GreenShip III being the most advanced. Different grades correspond to distinct EEDI value requirements, denoted as EEDI(I), EEDI(II), and EEDI(III) respectively. The achieved EEDI value must be lower than the required EEDI value, expressed as follows:

$$\text{Attained}\, EEDI \leq \text{Required}\, EEDI = (1 - X/100) \times RLV \quad (3.64)$$

where RLV represents the ship's reference line value, while X is the reduction coefficient used to determine the necessary EEDI corresponding to different energy efficiency requirements.

In this scenario, EEDI(I) is considered the required EEDI. Additionally, the EEDI refers to the value of the full-scale ship.

(4) Uncertain quantitative selection

During the actual voyage, various unavoidable disturbances such as wind and waves cause fluctuations in the operational speed and draft of the ship, deviating from the hydrostatic conditions. With the widespread adoption of the Safety of Life at Sea (SOLAS) Convention, nearly all ships are equipped with Automatic Identification System (AIS) equipment, enabling accurate real-time monitoring of a ship's dynamic position and data via electronic charts on the network. When monitoring ships along predetermined routes, the speed and draft approximately adhere to the following relationship:

$$V_{actual} = V_k + v \quad (3.65)$$

$$T_{actual} = T + t \quad (3.66)$$

where V_{actual} and T_{actual} represent the actual speed and draft, V_k and T denote the design speed and draft, and v and t signify the perturbations in speed and draft (as listed in Table 3.30).

(5) Optimise the algorithm.

The PSO algorithm has been chosen for the optimisation process. In this endeavour, a particle swarm size of 60 and a total of 20 iterations are utilised.

Table 3.30 RBRDO uncertainties

Uncertainty quantity	Distribution type and parameters
v	Normal distribution, $\mu = 0, \sigma = 0.1V_k$
t	Normal distribution, $\mu = 0, \sigma = 0.005T$

(6) Optimise the process.

The optimisation process is shown in Fig. 3.20 and entails several steps:

(1) Determination of design variables and their respective ranges.
(2) Generation of a new KCS hull form using the RBF interpolation method, based on a set of design variables, which is then fed into the optimisation module.
(3) Selection of the type of optimisation. If DO is chosen, ship resistance is provided by CFD tools, and the EEDI is obtained using Eq. (3.62). Otherwise, the procedure proceeds to the uncertainty analysis module.
(4) Utilisation of the IPCM in the uncertainty analysis module to select collocation points for uncertain variables.
(5) Calculation of the objective and constraint using the PCE method, leading to the derivation of polynomial coefficients.
(6) Calculation of the mean and standard deviation of the objective function from the obtained coefficients. Simultaneously, the failure probability of the constraint is determined using the MEM and transmitted to the optimiser.

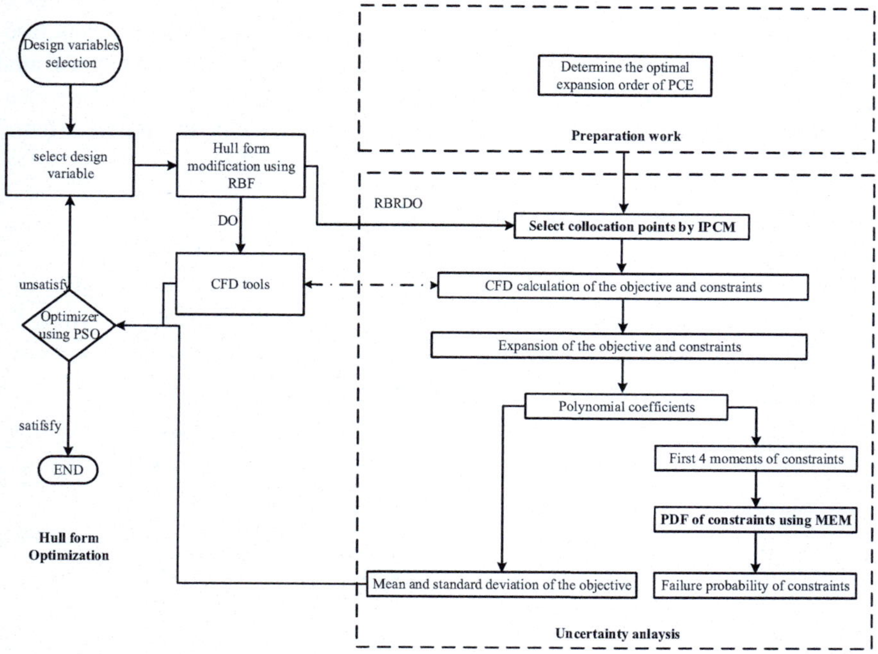

Fig. 3.20 Flow chart of RBRDO

(7) Assessment by the optimiser to determine whether the entire optimisation process meets the specified iteration requirements. If the process reaches the designated iteration number, the optimisation process concludes. If not, steps 3 to 7 are reiterated.

4. Optimise results and validations

The 5th PCE expansion is utilised for uncertainty analysis in the RBRDO and validated through the MC method with 105 sample points. The optimal solutions for both DO and RBRDO are presented in Table 3.31.

The table illustrates that in the case of RBRDO, the outcomes derived from both the PCE and MC methods exhibit close alignment, thereby reaffirming the viability of employing the PCE method for RBRDO. Additionally, these calculations were conducted on the same workstation. Conversely, for DO, where there is no requirement for uncertainty analysis, this method necessitates the least amount of time. In contrast, for RBRDO, while the MC method utilises an approximate model to mitigate computational demands, the PCE method entails solving time-consuming nonlinear equations to determine the probability of failure. Despite this, the overall time required for the MC method, encompassing both the computational time for CFD sampling and the optimisation time utilising the meta-model, remains approximately twice that of the total calculation time using the PCE method.

For geometric constraints, all variations are less than 0.5%, indicating minimal changes compared to the original hull design. Regarding reliability constraints, the PDF curve of EEDI is depicted in Fig. 3.21. The line representing EEDI(I) serves as the constraint boundary, delineating the failure area, which is defined by the intersection of the EEDI(I) line (shown as a dot-dashed line perpendicular to the horizontal axis) and the region bounded by the horizontal axis and the PDF curve. This visually illustrates the failure probabilities of the three schemes. Notably, RBRDO based

Table 3.31 Optimal solutions for DO and RBRDO

Item		Initial	DO	RBRDO	
				PCE	MC
Geometric constraints	$\lvert(\nabla - \nabla_0)/\nabla_0\rvert$	–	0.43%	0.02%	0.09%
	$\lvert(X_B - X_{B0})/X_{B0}\rvert$	–	0.12%	0.09%	0.13%
Reliability constraints	$P_f(EEDI < EEDI(I))$	21.5%	18.6%	4.8%	4.9%
Goals	$EEDI$	14.70	14.35	14.42	14.39
	Reduction	–	−2.38%	−1.90%	−2.11%
	$\mu(EEDI)$	17.35	16.96	17.17	17.05
	Reduction	–	−2.25%	−1.04%	−1.73%
	$\sigma(EEDI)$	4.54	4.43	4.39	4.36
	Reduction	–	−2.42%	−3.74%	−3.96%
Calculate time /h			14h	260h	515h
Calculate cost criteria	Using PSO algorithm to20 generations of time				

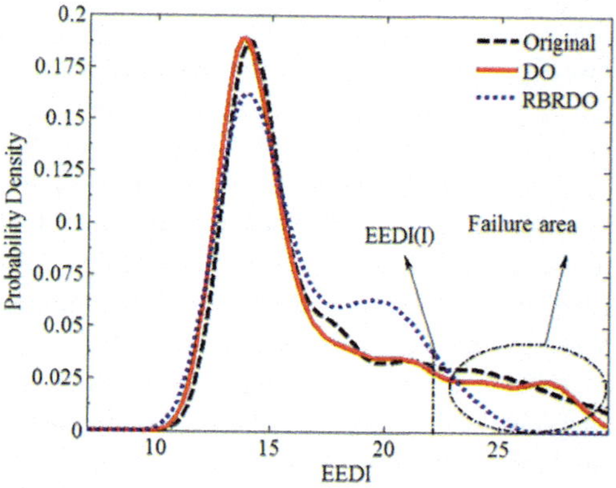

Fig. 3.21 Comparison of the probability density curves of the EEDI

on the PCE method exhibits the smallest regional area and thus the lowest failure probability. As indicated in Table 3.31, considering the impact of uncertainties, both the original and DO cases exhibit constraint violations with failure probabilities of 21.5% and 18.6% respectively, exceeding the failure standard of 5% and leading to scheme failure. This underscores the imperative of reliability analysis. Following optimisation, the failure probability of the RBRDO case is reduced to 4.8%, representing a significant enhancement in reliability and a substantial deviation from the failure threshold.

Furthermore, the EEDI values of both the DO and RBRDO cases decreased by approximately 2.38 and 1.90% respectively compared to the initial schemes. Although the results of the RBRDO scheme marginally lag behind those of the deterministic scheme, the standard deviation of EEDI reduced by up to 3.74%, surpassing that of DO, which decreased by 2.42%. Hence, it can be deduced that the RBRDO approach demonstrates greater adaptability and robustness in the face of environmental uncertainties. Put differently, in real-world sailing conditions, the RBRDO scheme exhibits lower EEDI values and enhanced navigational capability.

Moreover, the results presented in Table 3.31 suggest that the RBRDO solution appears to be primarily influenced by the design reliability rather than its robustness. Specifically, the design reliability is enforced as a stringent yet obligatory constraint, where the failure probability of the original and DO cases must be reduced by approximately 15% to satisfy reliability requirements ($P_f \leq 5\%$). Consequently, the enhancement of the reliability of the RBRDO solution appears considerably more pronounced than its robustness. Essentially, if both design reliability and robustness are designated as objectives, the improvement in robustness should outweigh that achieved by the current RBRDO solution. However, the reliability will also be enhanced, albeit with a less substantial reduction in failure probability (less than 5%), as $P_f \leq 5\%$ ceases to be a mandatory constraint. In summary, to prevent an imbalance between robustness and reliability requirements, if a higher level of robustness is sought, it may be necessary to relax the reliability constraint.

As illustrated in Fig. 3.22, the pattern alteration in the DO scheme appears more pronounced compared to the RBRDO scheme. Examining the transverse profile diagram, it is evident that in the RBRDO case, barring the bow area, the hull shape closely resembles that of the original design. Conversely, in the DO case, near the ship's bow and close to the baseline, there is a noticeable tendency towards "expansion" in the hull form, indicating that a bulbous bow with increased girth leads to better resistance reduction. Additionally, in the region between the bow and the midship, the hull shape exhibits a "shrinkage" trend near the baseline, while displaying an "expansion" trend near the waterline. Consequently, the bow becomes slimmer, resulting in a decrease in the inlet flow angle and an increase in the inlet flow length, both of which contribute to the reduction of wave-making resistance.

Furthermore, upon examining the hull profiles, it is notable that the bulb shapes of both the DO and RBRDO designs exhibit an upward and forward trend. This tendency could potentially result in beneficial wave interference, particularly at high speeds.

As indicated in Table 3.32, the viscous resistance coefficient (C_V) for both cases, encompassing the frictional resistance coefficient (C_F) and the viscous pressure resistance coefficient (C_{PV}), exhibited only a modest reduction. However, the decrease in the wave-making resistance coefficient (C_W) for the two optimal solutions amounted to 7.50 and 6.61%, respectively. This reduction is a significant contributing factor to the decrease in the total resistance coefficient (C_T), which was approximately 1.26 and 1.10%, consequently leading to a reduction in the ship's EEDI value.

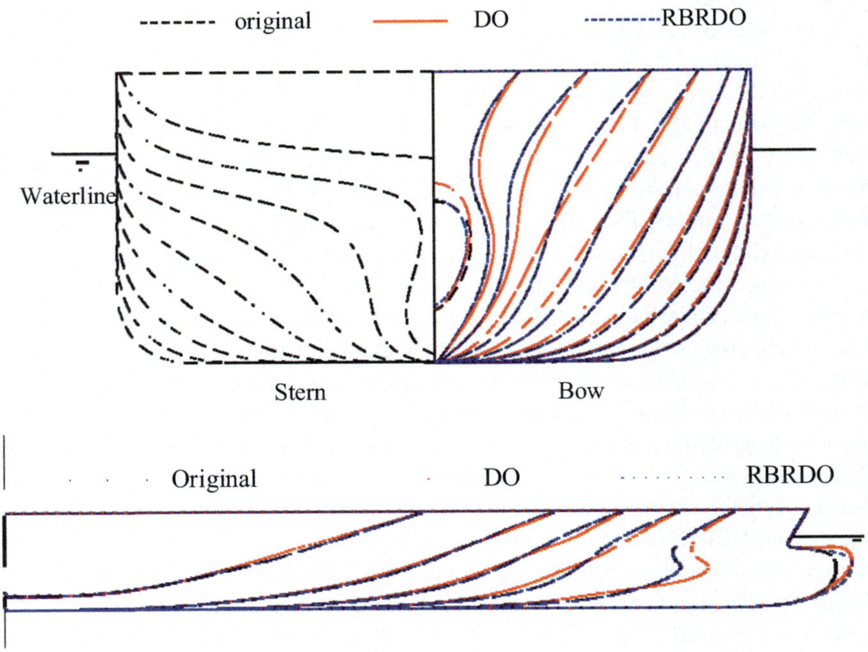

Fig. 3.22 Comparison of body plans and profiles between the original hull and the optimal hulls

Additionally, the wave profiles and contours of the original model and the two optimal models at the design speed are illustrated in Figs. 3.23 and 3.24, respectively. Observation of Fig. 3.23 reveals that the waves generated by the two optimal models at the design speed are smaller than those of the original model, with DO exhibiting a more pronounced reduction in wave contours compared to RBRDO. Furthermore, Fig. 3.24 demonstrates a significant decrease in wave amplitudes for the optimal models in comparison to the original model. Specifically, all wave troughs and peaks of the two optimal models are nearly lower than those of the original model, with the wave height of the RBRDO model falling between that of the DO model and the original model, resulting in a moderate reduction in resistance. It can be inferred that when accounting for the influence of uncertainty during the optimisation process, RBRDO must sacrifice some performance to ensure reliability and robustness, thereby meeting reliability demands and mitigating perturbations.

As depicted in Fig. 3.25, even minor alterations in hull form can induce variations in the surrounding pressure field, with the pressure exerted on the hull surface represented by distinct grayscale contours. Across the three schemes, the pressure distribution near the mid-ship remained largely consistent, while notable disparities were observed in the bow area. Specifically, the negative pressure area was markedly reduced on the two optimally modified bow surfaces, which contributes favourably to the reduction of pressure drag.

In conclusion, the modification to the hull shape in the case of DO is more extensive compared to RBRDO, resulting in a more pronounced reduction in resistance and, consequently, a more significant decrease in EEDI. However, since RBRDO does not

Table 3.32 The comparison of resistance coefficients

Scheme	C_F (%)	C_{PV} (%)	C_W (%)	C_T (%)
DO	−0.07	−0.50	−7.50	−1.26
RBRDO	−0.26	−0.31	−6.61	−1.23

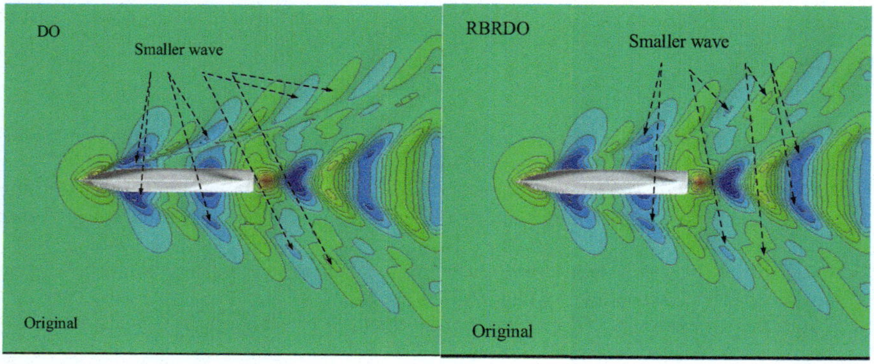

Fig. 3.23 Comparison of wave contours between the original and 2 optimal models

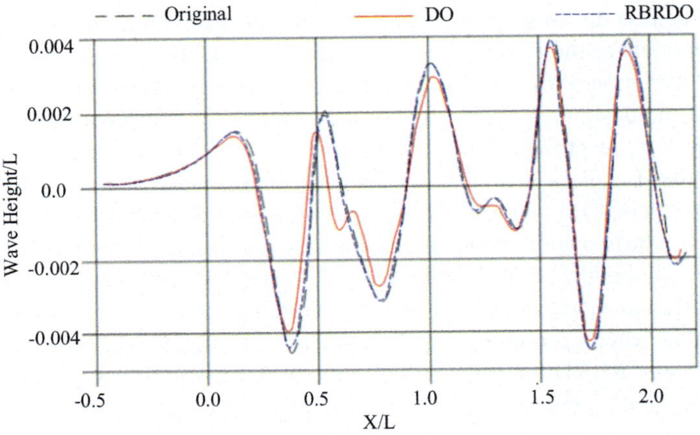

Fig. 3.24 Comparison of wave profiles between the original and 2 optimal models ($Y/L = -0.169$

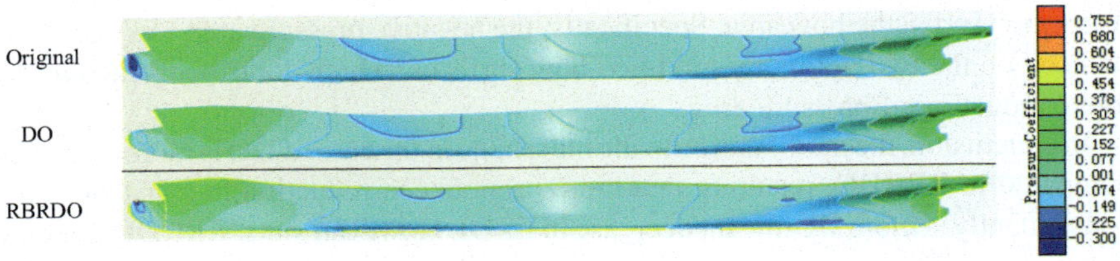

Fig. 3.25 Surface pressure coefficient contours for the original hull and the optimal hulls

entail a substantial alteration in hull shape, its objective exhibits lower perturbation, and the failure probability of the reliability bundle is also smaller than that of the DO scheme.

References

1. Wan X, Karniadakis GE. Multi-element generalized polynomial chaos for arbitrary probability measures. SIAM J Sci Comput. 2006;28:901–28.
2. Oladyshkin S, Class H, Helmig R, et al. A concept for datadriven uncertainty quantification and its application to carbon dioxide storage in geological formations. Adv Water Resour. 2011;34:1508–18.
3. Oladyshkin S, Nowak W. Data-driven uncertainty quantification using the arbitrary polynomial chaos expansion. Reliab Eng Syst Saf. 2012;106:179–90.
4. Cameron RH, Martin WT. The orthogonal development of nonlinear functionals in series of Fourier-Hermite functionals. Am J Math. 1947;48:385–92.
5. Askey R, Wilson JA. Some basic hypergeometric orthogonal polynomials that generalize Jacobi polynomials. Mem Am Math Soc. 1985; 54(319).

6. Shi L, Yang J, Zhang D, et al. Probabilistic collocation method for unconfined flow in heterogeneous media. J Hydrol. 2009;365(1):4–10.
7. Li W, Lu Z, Zhang D. Stochastic analysis of unsaturated flow with probabilistic collocation method. Water Resour Res. 2009;45(8):2263–89.
8. Huang SP, Jing P. Collocation based stochastic finite element method for reliability analysis of pile settlement. Chinese J Comp Mech. 2011;28:189–93.
9. Dianqing L, Zhou C, Chen Y, et al. Reliability analysis of slope using stochastic response surface method and code implementation. Chin J Rock Mech Eng. 2010;29(8):1513–23.
10. Kiureghian AD, Liu P. Structural reliability under incomplete probability information. J Eng Mech. 1986;112(1):85–104.
11. Isukapalli SS. Uncertainty analysis of transport-transformation models. Disser Theses Gradworks. 1999;57(1):31–2.
12. Jiang SH, Li DQ, Zhou CB. Optimal probabilistic collocation points for stochastic response surface method. Chinese J Comp Mech. 2012.
13. Sudret B. Global sensitivity analysis using polynomial chaos expansions. Reliab Eng Syst Saf. 2007;93(7):964–79.
14. Deman G, Konakli K, Sudret B, et al. Using sparse polynomial chaos expansions for the global sensitivity analysis of groundwater lifetime expectancy in a multi-layered hydrogeological model. Reliab Eng Syst Saf. 2016;147:156–69.
15. Fajraoui N, Ramasomanana F, Younes A, et al. Use of global sensitivity analysis and polynomial chaos expansion for interpretation of nonreactive transport experiments in laboratory-scale porous media. Water Resourc Res. 2011; 47(2).
16. Younes A, Mara TA, Fajraoui N, et al. Use of global sensitivity analysis to help assess unsaturated soil hydraulic parameters. Vadose Zone J. 2013; 12(1).
17. Sochala P, Maître OPL. Polynomial chaos expansion for subsurface flows with uncertain soil parameters. Adv Water Resour. 2013;62(62):139–54.
18. Ciriello V, Riva M, Cadini F, et al. Polynomial chaos expansion for global sensitivity analysis applied to a model of radionuclide migration in a randomly heterogeneous aquifer. Stoch Env Res Risk Assess. 2013;27(4):945–54.
19. Formaggia L, Imperiali I, Scotti A, et al. Global sensitivity analysis through polynomial chaos expansion of a; basin-scale geochemical compaction model. Comput Geosci. 2013;17(1):25–42.
20. Sobol IM. Theorems and examples on high dimensional model representation. Reliab Eng Syst Saf. 2003;79(2):187–93.
21. Jaynes ET. Information theory and statistical mechanics. Phys Rev. 1957;106(4):620–30.
22. Zhang X, Pandey MD. Structural reliability analysis based on the concepts of entropy, fractional moment and dimensional reduction method. Struct Saf. 2013;43(9):28–40.
23. Dai YS, Xie M, Long Q, et al. Uncertainty analysis in software reliability modeling by Bayesian analysis with maximum-entropy principle. IEEE Trans Software Eng. 2007;33(11):781–95.
24. Ovalle A, Fernandez J, Hably A, et al. An electric vehicle load management application of the mixed strategist dynamics and the maximum entropy principle. IEEE Trans Industr Electron. 2016;63(5):3060–71.
25. IMO-MEPC. 2014 Guidelines on the method of calculation of the attained energy efficiency design index (EEDI) for new ships. https:/www.imo.org/en/OurWork/Environment/Pages/Technical-and-Operational-Measures. Aspx [2014–2–15].
26. Cheng B, Pan WW. Ship design principle. Shanghai: Shanghai Jiao Tong University Press; 2007.

Chapter 4
Modelling of Epistemic Uncertainty

The challenge of epistemic uncertainty resides in the modelling methods. This chapter delves into extensive research on the modelling of cognitive uncertainty. The Akaike information criterion (AIC) is employed to identify the mixed distribution type of sparse variables, and the corresponding weights are assigned. The evidence theory is utilised to express interval variables, thus completing the cognitive uncertainty model.

4.1 Modelling of Sparse Uncertainty

In the context of sparse variables, parameters from six candidate distributions are estimated. Subsequently, the weights of these six candidate distributions are calculated using the AIC. Finally, parameter estimation is conducted on the mixed distributions comprised of these six distributions, thereby determining the final expression form to complete the modelling of sparse variables.

4.1.1 Weights of Sparse Variables

1. Parameters of Candidate Distributions

In the case of sparse variables, the PDF cannot be adequately represented by a single distribution. Hence, multiple distributions, weighted accordingly, are utilised to articulate sparse variables [1]. Six common distribution $\vartheta_k(k = 1, 2, \cdots, 6)$ are selected as candidate distributions. Assuming there are m point data $y_i(i = 1, 2, \cdots, m)$ and n interval data $[a_j, b_j](j = 1, 2, \cdots, n)$ for Y, with all data being independent, the likelihood function of each candidate distribution is:

Z. Liu et al., *Uncertainty-Based Ship Design Optimization*,
https://doi.org/10.1007/978-981-97-2973-9_4

$$L(\xi, \vartheta_k) \propto \left[\prod_{i=1}^{m} f_Y(y_i|\xi, \vartheta_k)\right]\left[\prod_{j=1}^{n}\left[F_Y(b_j|\xi, \vartheta_k) - F_Y(a_j|\xi, \vartheta_k)\right]\right] \tag{4.1}$$

The maximum likelihood estimator of ξ under the distribution type ϑ_k is obtained by maximising $L(\xi, \vartheta_k)$.

2. Weight Determination Based on AIC

AIC serves as a widely employed information criterion for evaluating the fitting performance of statistical models, particularly in model selection within statistical analysis. Generally, AIC is a weighted function considering both the unknown fitting accuracy and the number of parameters, defined as:

$$\mathrm{AIC}_k = 2\mathrm{num}_k - 2\ln(L_{\max}(\xi|\vartheta_k)),\ k = 1, 2, \cdots, 6 \tag{4.2}$$

where num_k is the number of unknown parameters in the candidate distribution ϑ_k, and $L_{\max}(\xi|\vartheta_k)$ is the maximum likelihood function of the distribution ϑ_k.

The model with the lowest AIC is generally regarded as the optimal model.

The AIC method is employed to compute the information loss of a sparse variable within a given distribution ϑ_k. This calculation assists in determining the most suitable distribution type for the sparse mixed uncertainty variable. For the candidate distribution $\vartheta_k (k = 1, 2, \cdots, 6)$, AIC_k is computed. For the kth distribution, the minimum probability formula of information loss is expressed as:

$$P_{\vartheta_k} = \exp\left(\frac{\mathrm{AIC}_{\min} - \mathrm{AIC}_k}{2}\right) \tag{4.3}$$

The distributions of $P_{\vartheta_k} \geq 0.1$ are selected to represent the sparse variables. The weights w_k are proportional to the probabilities P_{ϑ_k}, with the sum of weights equalling 1:

$$w_k = \frac{P_{\vartheta_k}}{\sum P_{\vartheta_k}} \tag{4.4}$$

4.1.2 Parameters of Sparse Variables

After establishing the weights of the six candidate distributions, the PDF of the sparse variables is represented as:

$$f_Y(Y|\xi) = \sum w_k f_Y(Y|\xi, \vartheta_k) \tag{4.5}$$

where ϑ_k is the distribution type, w_k is the corresponding weight, and $f_Y(Y|\boldsymbol{\xi}, \vartheta_k)$ is the PDF of the distribution ϑ_k.

$$L(\xi) \propto \left[\prod_{i=1}^{m} f_Y(y_i|\xi)\right]\left[\prod_{j=1}^{n}\left[F_Y(b_j|\xi) - F_Y(a_j|\xi)\right]\right] \tag{4.6}$$

The optimal distribution parameters ξ^* are computed for the selected distribution type by maximising $L(\xi)$.

4.1.3 Example

Utilising the example provided in Sect. 2.4 for further investigation, the uncertain parameters (E_1,E_2,V_{21},G_{12},t) are considered as sparse variables. Initially, the AIC values are computed for the five uncertain parameters, satisfying each of the six candidate distributions, as presented in Table 4.1. The corresponding weights are then delineated in Table 4.2.

Following the completion of weight calculations, the parameters that express the mixed distribution of the sparse variables are determined, as shown in Table 4.3.

From Table 4.1, it is evident that for all five uncertain parameters, the uniform distribution exhibits the smallest AIC value. Table 4.2 illustrates that the contribution

Table. 4.1 AIC of 6 candidate distributions

Distributions	E_1/GPa	E_2/GPa	V_{21}	G_{12}/GPa	t/mm
Normal	45.36	6.45	−42.02	−0.18	−0.56
Uniform	41.50	4.93	−43.93	−4.17	−4.99
Exponential	96.08	53.51	0.20	44.21	44.22
Weibull	46.71	7.53	−42.56	0.46	−2.95
Extreme value	45.17	7.73	−39.45	0.24	2.41
Lognormal	45.30	6.46	−41.78	−0.19	−0.26

Table. 4.2 Weights of 6 candidate distributions

Distributions	E_1/GPa	E_2/GPa	V_{21}/1	G_{12}/GPa	t/mm
Normal	0.095	0.191	0.164	0.092	0.069
Uniform	0.655	0.407	0.428	0.675	0.630
Exponential	0.000	0.000	0.000	0.000	0.000
Weibull	0.048	0.111	0.216	0.067	0.226
Extreme value	0.104	0.101	0.045	0.074	0.016
Lognormal	0.098	0.190	0.146	0.092	0.059

Table. 4.3 Parameters of sparse variables

Variables	Parameters	Normal	Uniform	Exponential	Weibull	Extreme value	Lognormal
E_1	a	130.23	126.90	130.62	131.96	133.72	5.53
	b	3.02	137.33	0.01	39.10	2.91	0.02
E_2	a	9.19	8.67	9.25	9.30	9.21	2.22
	b	0.21	9.73	0.11	32.15	0.20	0.03
V_{21}	a	0.33	0.30	0.32	0.33	0.34	−1.11
	b	0.01	0.35	3.09	40.00	0.01	0.02
G_{12}	a	5.12	4.85	5.13	5.11	5.11	1.63
	b	0.18	5.45	0.19	26.77	0.15	0.04
t	a	5.33	4.76	5.18	5.16	5.36	1.65
	b	0.16	5.33	0.19	30.95	0.18	0.04

of the exponential distribution to the mixed distribution is nearly zero. Specifically, for E_1, the optimal fitting distributions include the uniform and extreme value distributions; for E_2, the best fitting distributions encompass all five distributions except the exponential distribution; for V_{21}, the preferred distributions are normal, uniform, Weibull, and lognormal distributions; for G_{12}, the optimal fitting distribution is the uniform distribution; finally, for t, the most suitable distributions are the uniform and Weibull distributions.

4.2 Modelling and Propagation of Interval Variables

4.2.1 Evidence Theory

Evidence theory, a mathematical framework, determines belief and plausibility, quantifying uncertainty based on known evidence (information) for a proposition. Instead of assigning exact probabilities to propositions, it defines lower and upper bounds (intervals) on probabilities, particularly beneficial when uncertainty knowledge is limited. Information or evidence for measuring belief and plausibility originates from various sources, including experimental data, theoretical evidence, and expert opinions on parameters or event occurrences. Additionally, evidence can be amalgamated through fusion rules.

Let Ω be a generic set denoting all possible states of the system under consideration, whose power set 2^{Ω} represents propositions related to the system's actual state. Evidence theory assigns the BPA to each element of the set through the function m: $2^{\Omega} \to [0, 1]$, which adheres to the following three properties:

$$m(A) \geq 0, \forall A \in 2^{\Omega} \tag{4.7}$$

$$m(\phi) = 0 \tag{4.8}$$

$$\sum_{A \in 2^{\Omega}} m(A) = 1 \tag{4.9}$$

where the set $m(A) \geq 0$ is termed the focal element, while the BPA signifies the extent to which evidence supports an element of Ω belonging to the set A, akin to the PDF in probabilistic methods. When multiple interval variables are present, multiple BPA structures can be aggregated using fusion rules, akin to joint probability in probability theory. For instance, if there are interval variables Z_1 and Z_2, the joint BPA can be computed using Eq. (4.10).

$$S_Z(A_Z) = \begin{cases} S_{Z_1}(A_1)S_{Z_2}(A_2), & A_Z = A_1 \times A_2 \\ 0, & \text{others} \end{cases} \tag{4.10}$$

4.2.2 *Uncertainty Quantification for Interval Variables*

For proposition A, due to inadequate information, the level of support can be more accurately described by intervals, namely Bel and Pl, rather than a single probability value. The probability interval, comprising Bel and Pl measures, encompasses all potential cases of A without omission, indicating that the actual probability distribution of A lies between the upper and lower bound probabilities:

$$Bel \leq P(A) \leq Pl \tag{4.11}$$

From Fig. 4.1, the difference between the upper and lower bounds of A is $U = \text{Pl} - \text{Bel}$, which expresses the extent of epistemic uncertainty.

Bel represents the sum of BPA of evidence fully supporting proposition A, while Pl represents the sum of BPA fully or partially supporting proposition A. These measures

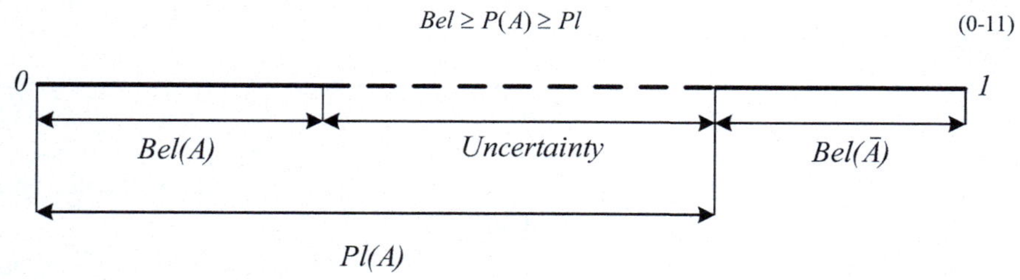

Fig. 4.1 Belief and plausibility

delineate the upper and lower bounds of the probability interval, characterising the uncertainty of proposition A.

$$Bel(A) = \sum_{B|B \subseteq A} m(B) \tag{4.12}$$

$$Bel(A) = \sum_{B|B \subseteq A} m(B) \tag{4.13}$$

Particularly, belief and plausibility are pivotal in reliability computations. As robust statistical variables X and sparse variables Y are represented by PDF, and interval variables Z are represented by evidence theory, the uncertainty of response cannot be quantified solely by determining probability values. Instead, it must be gauged using belief and plausibility, akin to probability intervals. According to evidence theory, the distribution of the response function g can be articulated by the cumulative belief function (CBF) and the cumulative plausibility function (CPF), defined as:

$$CBF(a) = Bel\{g < a\} \tag{4.14}$$

$$CPF(a) = Pl\{g < a\} \tag{4.15}$$

where, a denotes the limit state value, and Bel and Pl can be determined through reliability analysis. The CBF and CPF can be established by calculating Bel and Pl for different limit state values a.

In summary, evidence theory furnishes a broader approach to uncertainty modelling, allowing for a shift towards probabilistic approaches as additional information becomes available. This versatility in both describing and addressing uncertainty lays the groundwork for dealing with mixed uncertainty.

4.2.3 Numerical Example

To illustrate the computation of the Bel and Pl measures under evidence theory, we select the monotonic system performance function:

$$G = (x + y)^x \tag{4.16}$$

where x and y are evidence variables and their respective focal elements and corresponding BPA values are listed in Table 4.4.

Upon examination of Table 4.4, we derive the focal elements of response G along with their corresponding BPA values, as shown in Table 4.5.

Table. 4.4 Focal elements and BPA of x and y

x	BPA	y	BPA
[2, 2.3]	0.35	[3.5, 4]	0.42
[2.3, 2.6]	0.40	[4, 5]	0.42
[2.6, 3]	0.25	[5, 6]	0.16

Table. 4.5 Focal elements and BPA of response G

Focal elements	BPA	Lower bound	Upper bound
G_1	0.147	30.250	68.942
G_2	0.147	36.000	96.748
G_3	0.056	49.000	129.981
G_4	0.168	57.001	135.149
G_5	0.168	68.942	195.036
G_6	0.064	96.748	268.965
G_7	0.105	110.118	343.000
G_8	0.105	135.149	512.000
G_9	0.050	195.036	729.000

Subsequently, the Bel and Pl measures of response G are determined using Formulas (4.12) and (4.13) for various limit values a, depicted in Fig. 4.2.

Belief and plausibility measures of response G represent the lower and upper bounds of probability, respectively, with the actual probability distribution lying between them. The probability interval, comprising belief and plausibility measures, encompasses all potential probability distributions of response G without omission. The disparity between belief and plausibility measures signifies the magnitude of

Fig. 4.2 CBF and CPF of the response G

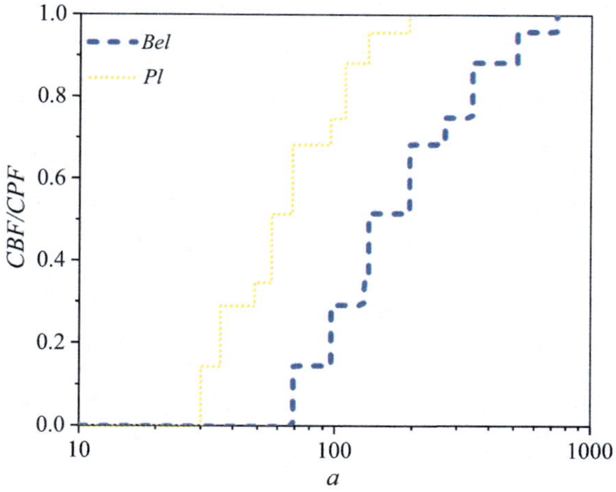

epistemic uncertainty in response *G*. As the epistemic uncertainty of the input parameter decreases, the gap between belief and plausibility measures gradually diminishes. If the epistemic uncertainty of the input parameter reaches zero, belief and plausibility measures coincide, indicating the presence of only aleatory uncertainty.

Reference

1. Nannapaneni S, Hu Z, Mahadevan S. Uncertainty quantification in reliability estimation with limit state surrogates[J]. Struct Multidiscip Optim. 2016; 54(6).

Chapter 5
Mixed Uncertainty Analysis and Propagation Method

In uncertainty modelling, when designers have access to sufficient available information, determining the exact PDF of uncertain parameters becomes feasible, ensuring the accuracy of uncertainty analysis. However, in many instances, the available information about variables is limited, making it challenging to employ probabilistic methods. In contrast, estimating the range of uncertain parameters is relatively straightforward. In such scenarios, non-probabilistic methods offer significant advantages in determining the expression of uncertain parameters with limited information. Substantial advancements have been recently made in both probabilistic and non-probabilistic modelling approaches. However, in certain cases, stochastic and cognitive uncertainties coexist. Many scholars have attempted to unify multiple types of uncertain parameters into a single uncertainty before conducting uncertainty analysis and propagation. Yet, this conversion lacks rigour and may introduce significant biases into the results. Therefore, the accurate performance of uncertainty analysis and propagation for systems under a mixture of stochastic and cognitive uncertainties is a topic warranting in-depth discussion.

5.1 Unified Analysis and Propagation Method of Mixed Uncertainty Based on Evidence Theory

Evidence theory, functioning as a non-probabilistic modelling approach, tends to approximate probabilistic methods in describing and managing uncertainty as more information becomes accessible. This characteristic sets the groundwork for adopting this method for the analytical transfer of mixed uncertainty. Consequently, this section addresses strong statistical variables, sparse variables, and interval variables concurrently to formulate expressions of robustness and believability under mixed uncertainty using evidence theory. This endeavour aims to accomplish a unified analytical transfer of mixed uncertainty.

Z. Liu et al., *Uncertainty-Based Ship Design Optimization*,
https://doi.org/10.1007/978-981-97-2973-9_5

5.1.1 Calculation of Believability and Plausibility Under Mixed Certainty

Under the influence of mixed uncertainty, the believability and plausibility of the response function $g(X, Y, Z)$ in the failure domain $D\{g(X, Y, Z) < a\}$ are calculated by considering the strong statistical variable X, sparse variable Y, and interval variable Z.

For the interval variable Z, its distribution function comprises N_c focal elements, expressed as c_k $(1 \le k \le N_c)$. To determine the believability and plausibility of failure, X and Y are set at their expected values, $X=\overline{X}$ and $Y = \overline{Y}$, respectively, to calculate the conditional failure believability at this point:

$$Bel(D|X=\overline{X}, Y = \overline{Y}) = \sum_{k=1}^{N_c} m(c_k)\delta_{Bel}(c_k, \overline{X}, \overline{Y}) \tag{5.1}$$

$$\delta_{Bel}(c_k, \overline{X}, \overline{Y})=\begin{cases} 1, \ \forall Z \in c_k, \ g(\overline{X}, \overline{Y}, Z) < a \\ 0, \ \text{others} \end{cases} \tag{5.2}$$

By integrating the conditional failure believability, the failure believability can be obtained:

$$\begin{aligned} Bel(D) &= \int_X \int_Y \rho(\overline{X})\rho(\overline{Y})Bel(D|X=\overline{X}, Y = \overline{Y})d\overline{Y}d\overline{X} \\ &= \int_X \int_Y \rho(\overline{X})\rho(\overline{Y})\left[\sum_{k=1}^{N_c} m(c_k)\delta_{Bel}(c_k, \overline{X}, \overline{Y})\right]d\overline{Y}d\overline{X} \\ &= \sum_{k=1}^{N_c}\left[m(c_k)\left(\int_X \int_Y \rho(\overline{X})\rho(\overline{Y})\delta_{Bel}(c_k, \overline{X}, \overline{Y})d\overline{Y}d\overline{X}\right)\right] \end{aligned} \tag{5.3}$$

Here, $Bel(c_k)=\int_X \int_Y \rho(\overline{X})\rho(\overline{Y})\delta_{Bel}(c_k, \overline{X}, \overline{Y})d\overline{Y}d\overline{X}$ represents the focal element c_k of sub-credibility, indicating the integration of the strong statistical variable X and sparse variable Y in their respective probability spaces. For any point $(\boldsymbol{X}, \boldsymbol{Y})$ in this region, the interval variable Z assumes any value in focal element c_k with $g(\boldsymbol{X}, \boldsymbol{Y}, \boldsymbol{Z}) < a$. Thus, $Bel(c_k)$ can be transformed to:

$$\begin{aligned} Bel(c_k) &= \Pr\{\boldsymbol{X}, \boldsymbol{Y}|\forall \boldsymbol{Z} \in c_k, g(\boldsymbol{X}, \boldsymbol{Y}, \boldsymbol{Z}) < a\} \\ &= \Pr\{g_{\max}(\boldsymbol{X}, \boldsymbol{Y}, \boldsymbol{Z}) < a|\boldsymbol{Z} \in c_k\} \end{aligned} \tag{5.4}$$

Subsequently, the failure believability is determined by:

$$Bel(D) = \sum_{k=1}^{N_c} m(c_k)Bel(c_k) \tag{5.5}$$

Similarly, the failure plausibility is given by:

$$Pl(D) = \sum_{k=1}^{N_c} m(c_k)Pl(c_k) \tag{5.6}$$

where

$$\begin{aligned} Pl(c_k) &= \Pr\{\boldsymbol{X}, \boldsymbol{Y} | \exists \boldsymbol{Z} \in c_k, g(\boldsymbol{X}, \boldsymbol{Y}, \boldsymbol{Z}) < a\} \\ &= \Pr\{g_{\min}(\boldsymbol{X}, \boldsymbol{Y}, \boldsymbol{Z}) < a | \boldsymbol{Z} \in c_k\} \end{aligned} \tag{5.7}$$

5.1.2 *Robustness and Reliability Expressions Under Mixed Uncertainty*

For reliability, the constraint function $g(\boldsymbol{X}, \boldsymbol{Y}, \boldsymbol{Z})$ contains three uncertainties, where X is a strong statistical variable expressed by a determined single probability distribution; Y is a sparse variable, expressed by the sum of multiple weighted probability distributions; and Z is an interval variable, represented by multiple focal element intervals $[\eta_j, \eta_{j+1}]$ (j=1, 2, $\cdots$, N_c), along with the corresponding BPA values for these focal elements.

Therefore, the failure probability p_f is formulated as:

$$\begin{aligned} p_f &= \Pr\{g(\boldsymbol{X}, \boldsymbol{Y}, \boldsymbol{Z}) < a\} \\ &= \sum_{j=1}^{N_c} \Pr\{g(\boldsymbol{X}, \boldsymbol{Y}, \boldsymbol{Z}) < a | \boldsymbol{Z} \in [\eta_j, \eta_{j+1}]\} \Pr\{\boldsymbol{Z} \in [\eta_j, \eta_{j+1}]\} \\ &= \sum_{j=1}^{N_c} p_{f\_j} \Pr\{\boldsymbol{Z} \in [\eta_j, \eta_{j+1}]\} \end{aligned} \tag{5.8}$$

where $p_{f-j} = \Pr\{g(X, Y, Z) < a | Z \in [\eta_j, \eta_{j+1}]\}$.

Let p_{f-j}^L and p_{f-j}^U represent the minimum and maximum values of p_{f-j}, respectively. To reduce computational costs, a Kriging model about the constraint function g is established. The solution is obtained through a sampling method, $Z \in [\eta_j, \eta_{j+1}]$ samples zn sample points, and computing the state function g under the samples using the Kriging model. Under each sample of Z, sampling X and Y again provides the minimum and maximum values of g, Min_{j-q} and $\mathrm{Max}_{j-q}(q=1, 2, \cdots, zn)$ (formula). Consequently, p_{f-j}^{L} and p_{f-j}^{U} can be expressed as:

$$p_{f-j}^{L} = \frac{\mathrm{Num}(\mathrm{Max}_{j-q} < a)}{zn} \tag{5.9}$$

$$p_{f-j}^{U} = \frac{\mathrm{Num}(\mathrm{Min}_{j-q} < a)}{zn} \tag{5.10}$$

After computing p_{f-j}^{L} and p_{f-j}^{U}, the believability and plausibility of failure probability can be expressed as:

$$Bel(g) = p_f^L = \sum_{j=1}^{N_c} p_{f-j}^{L} \Pr\{Z \in [\eta_j, \eta_{j+1}]\} \tag{5.11}$$

$$Pl(g) = p_f^U = \sum_{j=1}^{N_c} p_{f-j}^{U} \Pr\{Z \in [\eta_j, \eta_{j+1}]\} \tag{5.12}$$

The expectation and standard deviation of the response for robustness considerations differ slightly under mixed uncertainty. In this scenario, the expected EV of the response function, $f(\boldsymbol{X}, \boldsymbol{Y}(\xi), \boldsymbol{Z})$, also becomes an interval $[\mathrm{EV}_{\min}, \mathrm{EV}_{\max}]$:

$$\begin{cases} \mathrm{EV}_{\min} = \frac{1}{n_{XY}} \sum_{\eta=1}^{n_{XY}} \inf f_{\eta,w} \\ \mathrm{EV}_{\max} = \frac{1}{n_{XY}} \sum_{\eta=1}^{n_{XY}} \sup f_{\eta,w} \end{cases} \tag{5.13}$$

Its interval width is:

$$\Delta \mathrm{EV} = \mathrm{EV}_{\max} - \mathrm{EV}_{\min} \tag{5.14}$$

It can be deduced that when the interval variable degenerates to a random variable, $\Delta \mathrm{EV} = 0$.

For the expectation, similarly, the midpoint of the interval for the expectation of the response function, $f(\boldsymbol{X}, \boldsymbol{Y}(\xi), \boldsymbol{Z})$, is defined as:

$$\mathrm{EV}_{\mathrm{mid}} = \frac{1}{2}(\mathrm{EV}_{\max} + \mathrm{EV}_{\min}) \tag{5.15}$$

For the number η times sampling, the interval midpoint of $f(\boldsymbol{X}, \boldsymbol{Y}(\xi), \boldsymbol{Z})$ is:

$$(f_\eta)_{\mathrm{mid}} = \frac{1}{2}\left(\inf f_\eta + \sup f_\eta\right) \tag{5.16}$$

For each sample of Z, when sampling X and Y η_{XY} times, the mean square error of all f_η interval midpoints relative to $\mathrm{EV}_{\mathrm{mid}}$ is defined to characterise the degree of $f(\boldsymbol{X}, \boldsymbol{Y}(\xi), \boldsymbol{Z})$ deviation:

$$\mathrm{STD} = \sqrt{\frac{1}{\eta_{XY} - 1} \sum_{\eta=1}^{\eta_{XY}} \left[(g_\eta)_{\mathrm{mid}} - \mathrm{EV}_{\mathrm{mid}} \right]^2} \tag{5.17}$$

5.1.3 *Calculation Flow of the Unified Analysis and Propagation Method for Mixed Uncertainty Based on Evidence Theory*

Let X be the strong statistical variable, expressed by the PDF of the determined parameters, Y the sparse variable, the weight of each probability distribution is obtained by AIC, and Z the interval variable, represented by several subintervals with BPA values, with the distribution parameter ξ obtained through MLE.

In terms of reliability analysis, three uncertain system constraint functions, $g(\boldsymbol{X}, \boldsymbol{Y}(\xi), \boldsymbol{Z})$, are considered. Using the sampling method, the minimum and maximum values of the constraint function g, Min_{j-q} and Max_{j-q}, are calculated for each interval $Z \in [\eta_j, \eta_{j+1}]$. Subsequently, p_{f-j}^L and p_{f-j}^U are obtained, and Eqs. (5.11) and (5.12) are utilised to determine the believability and plausibility of constraint failure probability.

Similarly, for robustness assessment, the upper and lower limits of the objective function $f(\boldsymbol{X}, \boldsymbol{Y}(\xi), \boldsymbol{Z})$ can be determined using Eq. (5.13). Then, Eqs. (5.15) and (5.17) are employed to calculate the mean and standard deviation indicators of the robustness requirement, representing the midpoint and radius of the objective function's interval.

5.1.4 *Numerical Example*

1. Mathematical function

To verify the effectiveness of the unified analysis and propagation method of uncertainty based on evidence theory, function M is selected for testing:

$$M = (x + y)^z \tag{5.18}$$

where x is a strong statistical variable, y is a sparse variable, and z is an interval variable.

The strong statistical variable x follows a normal distribution $x \sim N(6, 0.5^2)$. Table 5.1 outlines the interval variable z and its corresponding BPA values for the point data {3.8, 4.1, 5.7} of sparse variable y, and the interval data [3.5, 4], [3.9, 4.1], [5, 6].

From Table 2.1, six competing distribution types ϑ are selected to determine the distribution parameters ξ for y. The MLE method is employed to solve the optimal parameters of the distribution, as presented in Table 5.2. Additionally, Fig. 5.1 illustrates the PDF curves under the six pending distributions and one mixed distribution for the sparse variable y.

Subsequently, employing the unified analysis and propagation method of uncertainty based on evidence theory, the uncertainty of the function M is estimated by considering three types of uncertainties separately. The results are then compared with those obtained through MC method verification. Believability and plausibility curves of $M = (x + y)^z$ were computed under various reliability limits C, as depicted

Table 5.1 The interval of interval variable z and its corresponding BPA value

Item	[2.00, 2.15]	[2.15, 2.50]	[2.30, 2.50]	[2.50, 2.60]	[2.60, 2.85]	[2.85, 3.00]
BPA	0.10	0.25	0.25	0.15	0.20	0.05

Table 5.2 Optimal parameters of the pending distribution

Distribution types	ξ_1	ξ_2	$Ln(L_{\max})$	AIC	Weight
Normal	4.515	0.868	−12.771	29.542	0.024
Uniform	3.500	6.000	−9.163	22.326	0.867
Exponential	4.515	0.221	−25.074	52.148	0.000
Weibull	4.884	5.578	−13.079	30.157	0.017
Extreme value	4.109	0.651	−11.939	27.878	0.054
Lognormal	1.490	0.187	−12.293	28.587	0.038

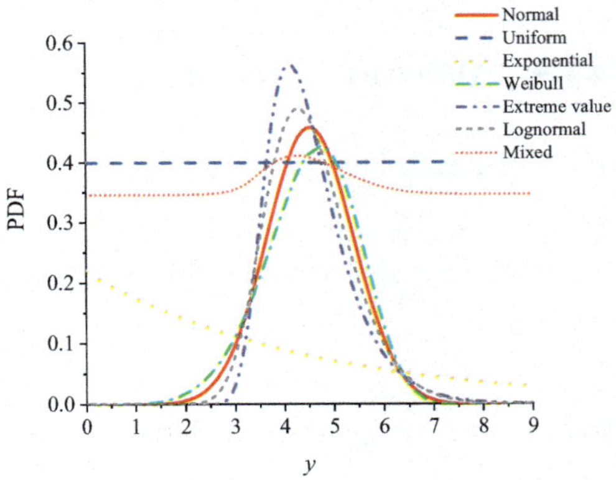

Fig. 5.1 PDF curve under 6 pending distributions and 1 mixed distribution for sparse variable y

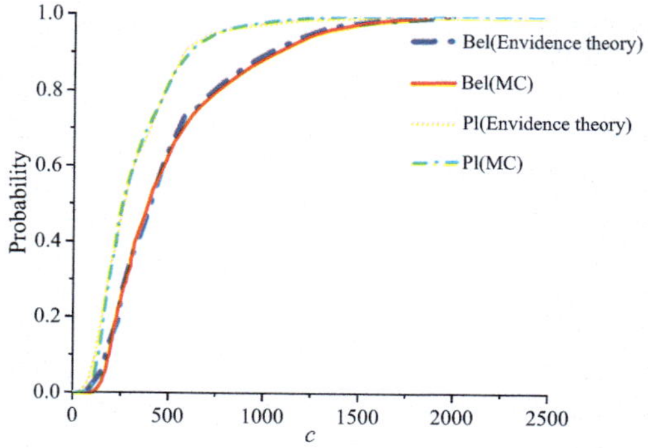

Fig. 5.2 Believability and plausibility curves for $M = (x + y)^z$

Table 5.3 Comparison of uncertain results of performance function M in limit state $c = 1000$

Item	Unified framework	MC method	Error/%
Believability	0.896	0.875	2.4
Plausibility	0.976	0.982	0.6

in Fig. 5.2. Table 5.3 displays the uncertainty result pairs of the performance function M under the limit state $c = 1000$. Notably, compared with the results obtained through the MC method, the mixed uncertainty analysis and propagation method based on evidence theory yields accurate uncertainty analysis results. Thus, the effectiveness of this method has been preliminarily verified.

2. Crank slider mechanism

To validate the efficacy of the unified analysis and propagation method based on evidence theory, the crank slider mechanism [1] shown in Fig. 5.3 is selected for examination. The length of the crank x_1, the length of the connecting rod x_2, and the external force x_3 are all strong statistical variables. The specific values for these variables are detailed in Table 5.4. Additionally, the Young's modulus of the connecting rod y_1 and its yield limit y_2 are considered sparse variables, with available data of y represented by point data {195, 200, 204} (GPa) and interval data [180, 197],[199, 209], [210, 212] (GPa). y_2 is the point data {280, 290, 298} (MPa) and interval data [270, 282], [288, 292], [299, 313](MPa). The friction coefficient z_1 and offset z_2 are interval variables, and their subintervals and BPA are outlined in Table 5.5. Moreover, the internal and external diameters of the connecting rod, d_1 and d_2, are 25 and 60 mm, respectively.

The function describing the entire system is the difference between the critical load and the axial load, i.e.:

$$G(\boldsymbol{X}, \boldsymbol{Y}, \boldsymbol{Z}) = \frac{\pi^3 y_1 (d_2^4 - d_1^4)}{64 x_2^2} - \frac{x_3 (x_2 - x_1)}{\sqrt{(x_2 - x_1)^2 - z_2^2 - z_1 z_2}} \tag{5.19}$$

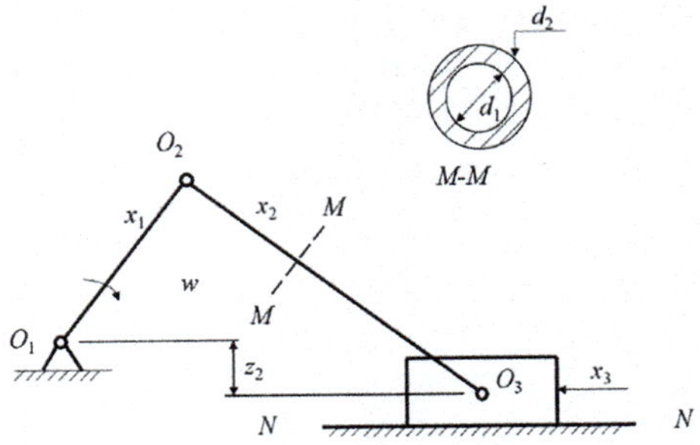

Fig. 5.3 Crank slider mechanism

Table 5.4 Strong statistical variable X

Variable	Mean value	Standard deviation	Distribution
x_1	100mm	0.01mm	Normal
x_2	400mm	0.01mm	Normal
x_3	250kN	25kN	Normal

Table 5.5 Subinterval and BPA of interval variable

Variable	Subinterval	BPA value
z_1	[100, 120]	0.2
	[120, 140]	0.4
	[140, 150]	0.4
z_2	[0.15, 0.18]	0.3
	[0.18, 0.23]	0.3
	[0.23,0.25]	0.4

To determine the distribution of Young's modulus, the AIC value of six alternative distributions is calculated. Consequently, Young's modulus is weighted and represented by the normal, Weibull, and extreme distributions, with weights of 0.415, 0.297, and 0.288, respectively.

The failure domain is defined by $F = \{\boldsymbol{X}, \boldsymbol{Y}, \boldsymbol{Z} | G(\boldsymbol{X}, \boldsymbol{Y}, \boldsymbol{Z}) < c\}$, where c denotes the limit state value. Employing the analysis and propagation method of mixed uncertainty based on evidence theory, the believability and plausibility under different reliability limits c are calculated and compared with the results of the MC method. The believability and plausibility curves of function G under various limit state values are depicted in Fig. 5.4.

It is evident that, within a range of $180 \leq c \leq 250$, the believability and plausibility of the functional function G calculated by the method proposed in this section are nearly identical to the results derived from the MC method. This finding reaffirms the feasibility of the unified analysis and propagation method for mixed uncertainty based on evidence theory.

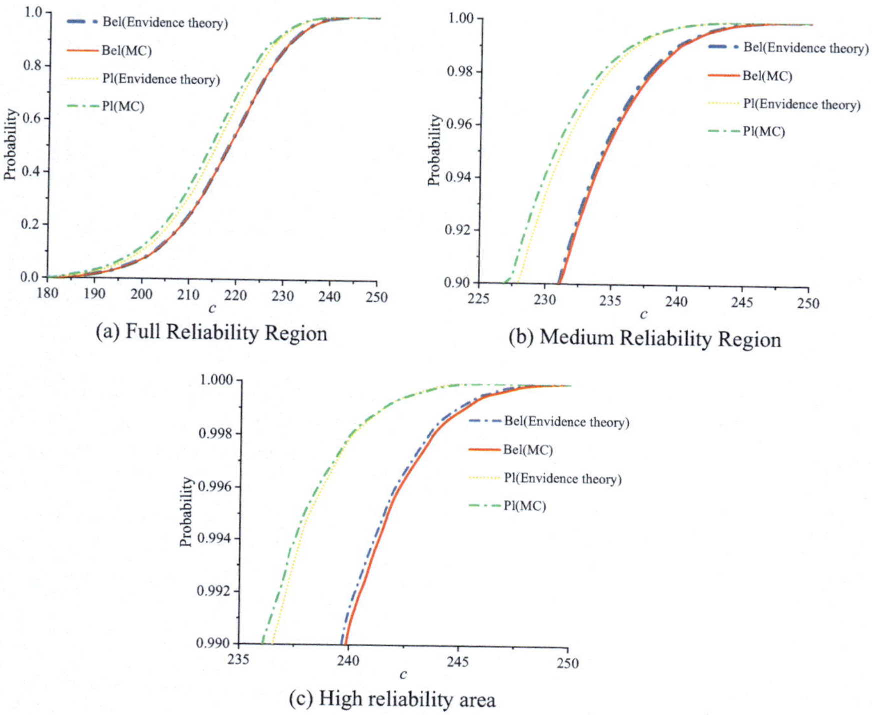

Fig. 5.4 The believability and plausibility curves of function G under different limit state values c

5.2 Unified Analysis and Propagation Method for Mixed Uncertainty Based on Evidence Theory Based on aPC Expansion

Previously, the unified analysis and propagation method for mixed uncertainty based on evidence theory was outlined. Numerical calculations demonstrated its effectiveness in analysing and propagating the uncertainty of system response quantities, irrespective of whether the response function is monotonic (e.g., M-function) or non-monotonic (such as the crank-slider mechanism), highlighting its adaptability. However, a notable drawback is the complexity of computation. Particularly, as the number of variable dimensions and focal elements increases, computational requirements escalate rapidly, posing a significant challenge in terms of computational cost and hindering its application in engineering contexts. Consequently, this section employs the aPC expansion method to undertake uncertainty analysis and propagation under mixed stochastic and cognitive influences.

5.2.1 aPC Expansion Method

1. Fundamental principle

For arbitrary PCE method, consider the probability space (Ω, A, Γ), where Ω is the sample space, A is the possible event in the sample space, and Γ is the probability measure. Then the model $Z(\xi)$ with input $\xi \in \Omega$ can be expressed as:

$$Z(\xi) \approx \sum_{i=0}^{d} c_i P^{(i)}(\xi) \tag{5.20}$$

where d is the expansion order, chosen to balance computational cost and model non-linearity; $\{P^{(0)}, P^{(1)}, \cdots, P^{(d)}\}$ are orthogonal polynomial bases, mutually orthogonal under probability measure Γ; and $c_i(i = \overline{0, d})$ determines the form of the PCE model.

The orthogonality of the orthogonal polynomial bases can be written as:

$$\int_{\xi \in \Omega} P^{(k)}(\xi) P^{(l)}(\xi) \mathrm{d}\Gamma(\xi) = \delta_{kl} \kappa_k \tag{5.21}$$

where δ_{kl} is the Kronecker function; and κ_k is an arbitrary constant representing the norm of the polynomial $P^{(k)}$, i.e.:

$$\kappa_k = \left\| P^{(k)}(\xi) \right\|^2 \tag{5.22}$$

When $\kappa_k \equiv 1(\forall k = \overline{0, d})$, the orthogonal polynomial bases $\{P^{(0)}, P^{(1)}, \cdots, P^{(d)}\}$ are orthogonal.

Unlike traditional the Wiener chaos expansion and generalised mixing expansion methods, aPC can be constructed under any probability metric Γ, allowing for the determination of orthogonal polynomial bases $\{P^{(0)}, P^{(1)}, \cdots, P^{(d)}\}$. The metric Γ can include discrete or continuous probability measures, which can be expressed through histograms or raw datasets, enabling the utilisation of any raw dataset to estimate widely approximated true probability measures:

$$\Gamma \approx \hat{\Gamma} = \frac{1}{N} \sum_{i=1}^{N} \delta(\xi - \xi_i) \tag{5.23}$$

where ξ_i is the available data set, N is the amount of data in the data set; and δ is the Dirac function.

Specifically, the expansion order is d determines any probability measure as an orthogonal polynomial base $\{P^{(0)}, P^{(1)}, \cdots, P^{(d)}\}$. By defining a polynomial of order k, $P^{(k)}(\xi)$, with random variable $\xi \in \Omega$, then:

$$P^{(k)}(\xi)=\sum_{i=0}^{k} p_i^{(k)}\xi^i, \quad k=\overline{0,d} \tag{5.24}$$

where $p_i^{(k)}$ is the coefficient of $P^{(k)}(\xi)$.

Substituting Eq. (5.24) into the orthogonal condition Eq. (5.21), $P^{(k)}$ is orthogonal with all polynomials $P^{(l)}(l \le k)$ smaller than its order, i.e.:

$$\begin{cases} \int\limits_{\xi\in\Omega} p_0^{(0)}\left[\sum_{i=0}^{k} p_i^{(k)}\xi^i\right] d\Gamma(\xi)=0 \\ \int\limits_{\xi\in\Omega} \left[\sum_{i=0}^{1} p_i^{(1)}\xi^i\right]\left[\sum_{i=0}^{k} p_i^{(k)}\xi^i\right] d\Gamma(\xi)=0 \\ \vdots \\ \int\limits_{\xi\in\Omega} \left[\sum_{i=0}^{k-1} p_i^{(k-1)}\xi^i\right]\left[\sum_{i=0}^{k} p_i^{(k)}\xi^i\right] d\Gamma(\xi)=0 \\ \int\limits_{\xi\in\Omega} \left[\sum_{i=0}^{k} p_i^{(k)}\xi^i\right]\left[\sum_{i=0}^{k} p_i^{(k)}\xi^i\right] d\Gamma(\xi)=\kappa_k \end{cases} \tag{5.25}$$

where κ_k is the number of norms.

By solving the above closed equations, the unique certainty coefficient $p_i^{(k)}$ can be obtained. The first equation can be decomposed as:

$$\int\limits_{\xi\in\Omega} p_0^{(0)}\left[\sum_{i=0}^{k} p_i^{(k)}\xi^i\right] \mathrm{d}\Gamma(\xi)=p_0^{(0)}\int\limits_{\xi\in\Omega}\sum_{i=0}^{k} p_i^{(k)}\xi^i\mathrm{d}\Gamma(\xi)=0 \tag{5.26}$$

From this, we can obtain:

$$\int\limits_{\xi\in\Omega}\sum_{i=0}^{k} p_i^{(k)}\xi^i\mathrm{d}\Gamma(\xi)=0 \tag{5.27}$$

The second equation can be decomposed as:

$$\begin{aligned}
&\int\limits_{\xi\in\Omega}\left[\sum_{i=0}^{1}p_i^{(1)}\xi^i\right]\left[\sum_{i=0}^{k}p_i^{(k)}\xi^i\right]\mathrm{d}\Gamma(\xi)\\
&=\int\limits_{\xi\in\Omega}\left(p_0^{(1)}\xi^0+p_1^{(1)}\xi^1\right)\left[\sum_{i=0}^{k}p_i^{(k)}\xi^i\right]\mathrm{d}\Gamma(\xi)\\
&=p_0^{(1)}\int\limits_{\xi\in\Omega}\sum_{i=0}^{k}p_i^{(k)}\xi^i\mathrm{d}\Gamma(\xi)+p_1^{(1)}\int\limits_{\xi\in\Omega}\sum_{i=0}^{k}p_i^{(k)}\xi^{i+1}\mathrm{d}\Gamma(\xi)\\
&=0
\end{aligned} \tag{5.28}$$

Substituting into Eq. (5.27), we get:

$$\int\limits_{\xi\in\Omega}\sum_{i=0}^{k}p_i^{(k)}\xi^{i+1}\mathrm{d}\Gamma(\xi)=0 \tag{5.29}$$

If each equation in the system of equations is similarly decomposed, the system can be rewritten by substituting the $1\sim k-1$ equations into the k equation, as follows:

$$\begin{cases}
\int\limits_{\xi\in\Omega}\sum_{i=0}^{k}p_i^{(k)}\xi^i\mathrm{d}\Gamma(\xi)=0\\
\int\limits_{\xi\in\Omega}\sum_{i=0}^{k}p_i^{(k)}\xi^{i+1}\mathrm{d}\Gamma(\xi)=0\\
\vdots\\
\int\limits_{\xi\in\Omega}\sum_{i=0}^{k}p_i^{(k)}\xi^{i+k-1}\mathrm{d}\Gamma(\xi)=0\\
\int\limits_{\xi\in\Omega}\sum_{i=0}^{k}p_i^{(k)}\xi^{i+k}\mathrm{d}\Gamma(\xi)=\dfrac{\kappa_k}{p_k^{(k)}}
\end{cases} \tag{5.30}$$

It is worth noting that if the origin moment of the random variable is:

$$\mu_k=\int\limits_{\xi\in\Omega}\xi^k\mathrm{d}\Gamma(\xi) \tag{5.31}$$

then the system can be rewritten in a form that contains only the moment information of the random variable. i.e.:

$$\begin{cases} \sum_{i=0}^{k} p_i^{(k)} \mu_i = 0 \\ \sum_{i=0}^{k} p_i^{(k)} \mu_{i+1} = 0 \\ \vdots \\ \sum_{i=0}^{k} p_i^{(k)} \mu_{i+k-1} = 0 \\ \sum_{i=0}^{k} p_i^{(k)} \mu_{i+k} = \frac{\kappa_k}{p_k^{(k)}} \end{cases} \tag{5.32}$$

For ease of calculation, Eq. (5.32) can be written in matrix form:

$$\begin{bmatrix} \mu_0 & \mu_1 & \cdots & \mu_k \\ \mu_1 & \mu_2 & \cdots & \mu_{k+1} \\ \vdots & \vdots & \vdots & \vdots \\ \mu_{k-1} & \mu_k & \cdots & \mu_{2k-1} \\ \mu_k & \mu_{k+1} & \cdots & \mu_{2k} \end{bmatrix} \begin{bmatrix} p_0^{(k)} \\ p_1^{(k)} \\ \vdots \\ p_{k-1}^{(k)} \\ p_k^{(k)} \end{bmatrix} = \begin{bmatrix} 0 \\ 0 \\ \vdots \\ 0 \\ \frac{\kappa_k}{p_k^{(k)}} \end{bmatrix} \tag{5.33}$$

The matrix on the left side of equation Eq. (5.33) is called the Hankel matrix of statistical moments. Lindsay [2] studied the properties of its determinants and concluded that an orthogonal polynomial basis of order d can be constructed with any random probability measure Γ and only if the Hankel matrix of moments is not a singular matrix.

Linearizing the system of equations and taking $p_k^{(k)}=1$, then Eq. (5.33) is transformed into:

$$\begin{bmatrix} \mu_0 & \mu_1 & \cdots & \mu_k \\ \mu_1 & \mu_2 & \cdots & \mu_{k+1} \\ \vdots & \vdots & \vdots & \vdots \\ \mu_{k-1} & \mu_k & \cdots & \mu_{2k-1} \\ 0 & 0 & \cdots & 1 \end{bmatrix} \begin{bmatrix} p_0^{(k)} \\ p_1^{(k)} \\ \vdots \\ p_{k-1}^{(k)} \\ p_k^{(k)} \end{bmatrix} = \begin{bmatrix} 0 \\ 0 \\ \vdots \\ 0 \\ 1 \end{bmatrix} \tag{5.34}$$

where

$$\mu_l = \int_{\xi_i} (\xi_i)^l \mathrm{d}\Gamma(\xi_i),\ l = 0, 1, \cdots, d \tag{5.35}$$

Decomposing Eq. (5.34) as:

$$M = \begin{bmatrix} H & B \\ C & D \end{bmatrix} \tag{5.36}$$

where $H = \begin{bmatrix} \mu_0 & \cdots & \mu_{k-1} \\ \vdots & \vdots & \vdots \\ \mu_{k-1} & \cdots & \mu_{2k-2} \end{bmatrix}$; $B = \begin{bmatrix} \mu_k \\ \vdots \\ \mu_{2k-1} \end{bmatrix}$; $C = \begin{bmatrix} 0 & \cdots & 0 \end{bmatrix}$; $D = [1]$.

To ensure that the transformed Eq. (5.34) can construct an orthogonal polynomial basis, matrix M must be a non-singular matrix.

Since D is invertible, the solution of the determinant of the partitioned matrix M is:

$$\det(M) = \det(D)\det(H - BD^{-1}C) \tag{5.37}$$

Because $\det(D)=1$, and $C = \begin{bmatrix} 0 & \cdots & 0 \end{bmatrix}$, we get:

$$\det(M) = \det(H) \tag{5.38}$$

The matrix H is the Hankel matrix. Karlin [3] shows that the determinant $\det(H)$ is zero when the rank of matrix $Rank(H) = k$, if and only if the number of sample points ξ of the distribution about $N \leq k$. Therefore, M is a non-singular matrix if and only if the number of ξ distributed sample points $N > k$, and the statistical moments of order $0 \sim 2k - 2$ are all determined.

Finally, by normalising the orthogonal polynomial, we obtain the orthogonal polynomial base $\{\varphi^k(\xi)\}$, $(k = \overline{0, d})$, for this probability measure Γ:

$$\varphi^{(k)}(\xi) = \frac{1}{\left\| P^{(k)} \right\|} \sum_{i=0}^{k} p_i^{(k)} \xi^i \tag{5.39}$$

In summary, utilising the raw statistical moments of the variables enables the construction of the Hankel matrix of the statistical moments to solve the orthogonal polynomial substrate, ultimately leading to the construction of the PCE model. Therefore, the determined statistical moments serve as a sufficient condition for constructing orthogonal polynomial bases. To streamline subsequent program writing and calculations, this book derives the moments of origin for the six distributions listed in Table 2.1, as shown in Appendix A.

2. Comparison of gPC and aPC

To contrast the aPC method presented in this section with the traditional gPC method, let us choose the simple nonlinear function:

$$Y_4 = \xi^6 \tag{5.40}$$

where ξ satisfies the Chi-square distribution.

This distribution does not belong to the Askey family, hence, if gPC is adopted, the random variable ξ needs to be transformed. Therefore, this book implements two PCE modelling methods: gPC, which transforms variables into normal distributions using Hermite polynomials, and another based directly on the Chi-square distribution. For the latter, higher-order statistical moments are directly solved, and optimal orthogonal polynomials are obtained using the aPC method for PCE modelling. Subsequently, the results of both methods are compared with the exact solution to assess their accuracy.

In the gPC model, the collocation points for solving the d-th order expansion are the roots of the $(d + 1)$-th order Hermite polynomial. However, the actual collocation points should be the roots of the orthogonal and optimal polynomial without any transformation.

As depicted in Fig. 5.5, the roots of the transformed polynomials deviate from the optimal polynomial roots, rendering them unsuitable for numerical integration. Moreover, the highly nonlinear transformation process leads to significant errors in the results of the PCE method. Ideally, the PCE model results should align with the exact solution when the expansion order matches the true order. Figure 5.6 illustrates that the results of the optimal polynomials with a 6th order expansion perfectly match the exact solution. Conversely, the transformed Hermite polynomial fails to converge to the exact solution, despite being of order 6, due to the use of non-optimal Gaussian integration points.

Fig. 5.5 Movement of roots after transformation

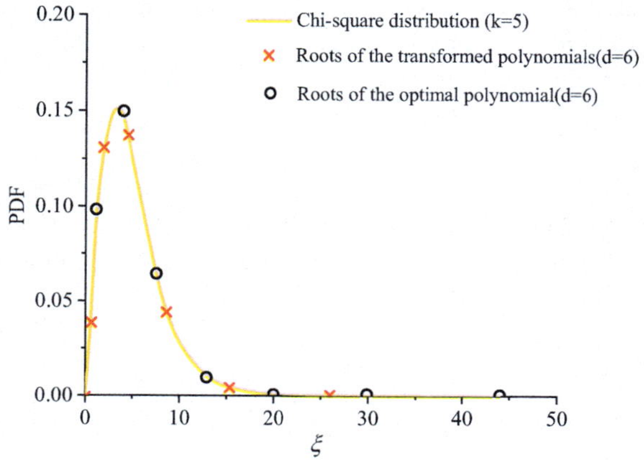

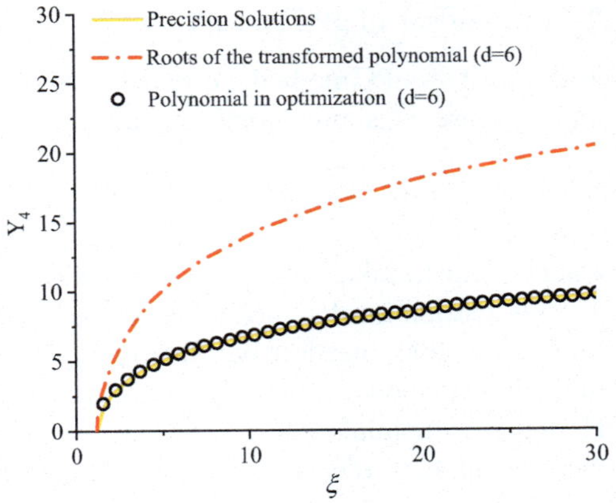

Fig. 5.6 Comparison of PCE modelling accuracy

It is evident that aPC eliminates the need for variable transformation when conducting PCE modelling of non-Askey family distributions. Moreover, its results are more accurate and applicable across a wider range compared to traditional gPC.

5.2.2 Metropolis-Hasting (M-H) Sampling

As demonstrated in Chap. 3, solving the PCE model coefficients requires sampling the uncertain variables and solving the coefficients through a regression method. For the strong statistical variable X, which is characterised by a single distribution expressing deterministic parameters, it can be sampled directly using the MC method or PCM. However, for the sparse variable Y, which is represented by the sum of multiple distributions with weights, traditional sampling methods encounter difficulties. Hence, the M-H sampling method is employed to sample the sparse variable Y in this section. This method, a form of Markov chain Monte Carlo (MCMC) sampling, finds widespread application in fields such as machine learning, deep learning, and natural language processing.

1. MCMC sampling

The MCMC sampling method consists of two MC methods. For the first MC method, if $f(x)$ integrates over interval $[a, b]$ (Fig. 5.7), a simple approach involves selecting a series of points in the interval $[a, b]$, where the PDF of x is $p(x)$. The definite integral can then be reformulated as:

$$I = \int_a^b f(x)\mathrm{d}x = \int_a^b \frac{f(x)}{p(x)} p(x)\mathrm{d}x \approx \frac{1}{n}\sum_{i=1}^{n} \frac{f(x_i)}{p(x_i)} \tag{5.41}$$

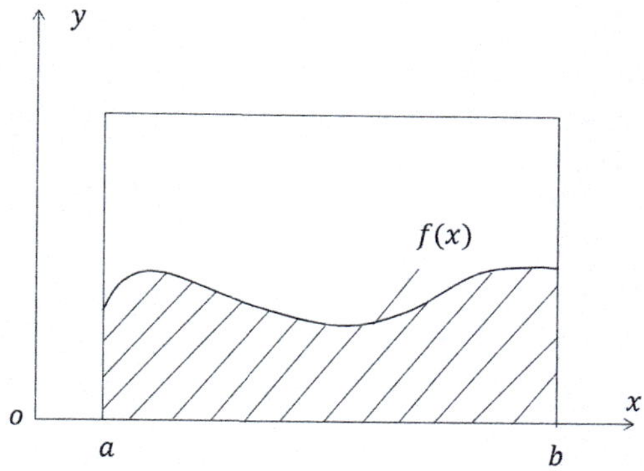

Fig. 5.7 Integration of $f(x)$

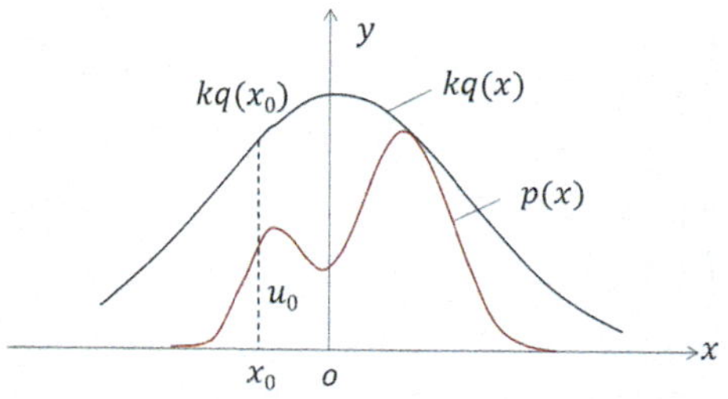

Fig. 5.8 Acceptance-rejection sampling schematic

For samples from the distribution corresponding to x, if x satisfies a common distribution, traditional sampling methods can readily obtain its sample set. However, for uncommon distributions, the accept-reject method may be employed to obtain samples from that distribution. For complex $p(x)$, a common PDF $q(x)$ is chosen, along with a constant k ensuring that $p(x)$ is always below $kq(x)$. The accept-reject sampling schematic is illustrated in Fig. 5.8.

First, sample to obtain a sample $q(x)$ from the mean distribution $[0, kq(x_0)]$. If u_0 falls between two curves, this sampling is rejected; otherwise, it is accepted, and the above steps are repeated until n sample points $x_0, x_1, \cdots, x_n$ are obtained.

However, accept-reject sampling has limitations. For instance, with high-dimensional complex non-conventional distributions, finding a suitable $q(x)$ and k can be challenging. The required sample size (or probability of acceptance) for sampling grows exponentially with the increase in spatial dimension. Hence, Markov chains are introduced to address this issue.

Markov chains assume that the probability of a state transfer at a given moment depends solely on its previous state. The Markov chain model converges to the final stable probability distribution through the state transfer matrix, independent of the initial state probability distribution. From this, the state transfer matrix corresponding to the final stable probability distribution of the model can be obtained by starting from an arbitrary sample of probability distributions, substituting into the state transfer matrix, and undergoing a sequence of transformations to obtain samples that conform to the corresponding stable probability distribution.

For a non-periodic Markov chain, if it possesses a state transfer matrix P and any two of its states are interconnected, meaning from any state one can reach any other state through a finite number of transfers, and there are no cases where the conditional probability is always zero and unreachable.

If $\lim_{n\to\infty} P_{ij}^n$ is independent of i, noting that $\lim_{n\to\infty} P_{ij}^n = \pi(j)$, then we have:

$$P = \begin{bmatrix} \pi(1) & \pi(2) & \cdots & \pi(j) & \cdots \\ \pi(1) & \pi(2) & \cdots & \pi(j) & \cdots \\ \vdots & \vdots & \ddots & \vdots & \vdots \\ \pi(1) & \pi(2) & \cdots & \pi(j) & \cdots \\ \cdots & \cdots & \cdots & \cdots & \cdots \end{bmatrix} \tag{5.42}$$

$$\pi(j) = \sum_{i=0}^{\infty} \pi(i) P_{ij} \tag{5.43}$$

and π is the only non-negative solution of $\pi = P\pi$, where $\pi = [\pi(1), \pi(2), \cdots, \pi(j), \cdots]$ and $\sum_{i=0}^{\infty} \pi(i) = 1$. π is the final smooth distribution of the model.

In summary, obtaining the Markov chain state transfer matrix corresponding to a certain smooth distribution simplifies sampling the sample set of this smooth distribution. The convergence property of the Markov chain is primarily determined by the transfer matrix P. Hence, constructing P such that the smooth distribution accurately matches the desired distribution $\pi(x)$ is crucial. Consequently, determining the Markov chain state transfer matrix involves ensuring meticulous smoothness conditions are met, i.e., the state transfer matrix P and the distribution $\pi(x)$ satisfy:

$$\pi(i) P(i, j) = \pi(j) P(j, i) \tag{5.44}$$

Then:

$$\sum_{i=0}^{\infty} \pi(i) P(i, j) = \sum_{i=0}^{\infty} \pi(j) P(j, i) = \pi(j) \sum_{i=0}^{\infty} P(j, i) = \pi(j) \tag{5.45}$$

Written in matrix form:

$$\pi P = \pi \tag{5.46}$$

From Eq. (5.46), it is evident that solving the state transfer matrix P is akin to finding a matrix whose probability distribution $\pi(x)$ satisfies meticulous smoothness conditions. However, this condition is quite stringent, and finding a random matrix Q that meets such requirements may be unfeasible:

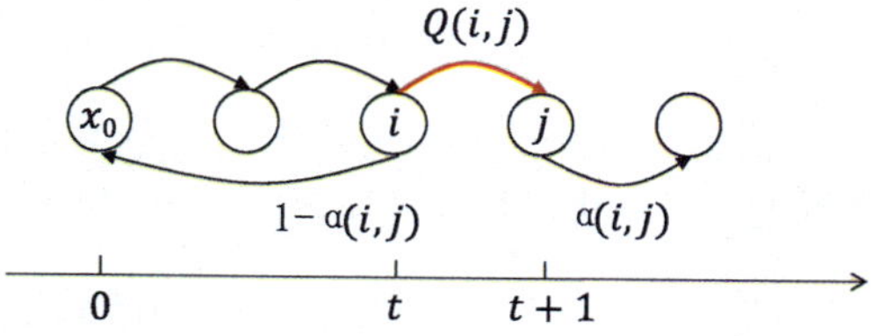

Fig. 5.9 Schematic representation of state transfer and acceptance probability of the Markov chain

$$\pi(i)Q(i,j) \neq \pi(j)Q(j,i) \tag{5.47}$$

To solve the above problem, $\alpha(i,j)$ is introduced, ensuring that the meticulous smoothness condition of Eq. (5.47) holds:

$$\pi(i)Q(i,j)\alpha(i,j) = \pi(j)Q(j,i)\alpha(j,i) \tag{5.48}$$

By symmetry, let:

$$\alpha(i,j) = \pi(j)Q(j,i) \tag{5.49}$$

$$\alpha(j,i)=\pi(i)Q(i,j) \tag{5.50}$$

Then Eq. (5.48) holds with:

$$\pi(i)\underbrace{Q(i,j)\alpha(i,j)}_{P(i,j)} = \pi(j)\underbrace{Q(j,i)\alpha(j,i)}_{P(j,i)} \tag{5.51}$$

A schematic of the Markov chain state transfer and acceptance probability is shown in Fig. 5.9.

In summary, akin to the acceptance-rejection sampling in the MC method, the MCMC sampling method achieves the target transfer matrix P through the acceptance-rejection probability of a common Markov chain state transfer matrix Q.

2. M-H sampling

Conventional MCMC sampling methods can prove highly inefficient for practical applications. This inefficiency arises from the potentially minute acceptance rate $\alpha(x_*, x_t)$, leading to the rejection of most sampled values during transfer. Consequently, achieving convergence to a smooth distribution becomes a slow process, necessitating n_1 to be exceedingly large, which could incur substantial costs in engineering applications.

M-H sampling, a type of MCMC sampling, offers a solution to the low acceptance rate issue encountered in traditional MCMC sampling. Initially proposed by Metropolis and refined by Hastings, the algorithm operates under the assumption that $\alpha(i,j)$=0.1 and $\alpha(j,i) = 0.2$, satisfying meticulous smoothness conditions, where:

$$\pi(i)Q(i,j) \times 0.1 = \pi(j)Q(j,i) \times 0.2 \tag{5.52}$$

At this point, the acceptance rate may be low, resulting in less efficient sampling. Expanding both sides of Eq. (5.52) by a factor of 5 yields:

$$\pi(i)Q(i,j) \times 0.5 = \pi(j)Q(j,i) \times 1 \tag{5.53}$$

This expansion enhances the acceptance rate without violating the meticulous smoothing condition. Thus, by scaling up both sides equally or scaling up the larger of the two to 1:

$$\alpha(i,j) = \min\{\frac{\pi(j)Q(j,i)}{\pi(i)Q(i,j)}, 1\} \tag{5.54}$$

With this slight modification, the M-H sampling process is as follows:

(1) Arbitrarily select a state transfer matrix Q, a smooth distribution, set a threshold for the number of state transfers n_1 and determine the required number of samples n_2.
(2) Sample from a simple probability distribution to initialise the Markov chain with an initial state value x_0.
(3) From $t = 0$ to $n_1+n_2 - 1$, x_* is obtained by sampling the conditional probability distribution $Q(x|x_t)$; the uniform distribution $u \sim uniform[0, 1][0, 1]$ is sampled. If $u < \alpha(x_*, x_t) = \pi(x_*)Q(x_*, x_t)$ then the state transfer $x_t \rightarrow x_*$ is accepted, i.e., $x_{t+1}=x_*$; otherwise, the transfer is not accepted and $x_{t+1}=x_*$. The sample set $(x_{n_1}, x_{n_1+1}, \cdots, x_{n_1+n_2-1})$ is the sample set for the desired smooth distribution.

Usually, the chosen matrix is symmetric, i.e., formula further simplifying the acceptance rate:

$$\alpha(i,j) = \min\{\frac{\pi(j)}{\pi(i)}, 1\} \tag{5.55}$$

3. Numerical examples

Since this chapter investigates the M-H method for sampling sparse variables, the target smooth distribution chosen is a mixture of distributions, represented as:

$$\begin{aligned}\pi(x) = {} & 0.2 * \text{Normal}(x, 0, 1) + 0.2 * \text{Uniform}(x, -1, 1) \\ & + 0.2 * \text{Exponential}(x, 1, 1) + 0.2 * \text{Weibull}(x, 2, 1) \\ & + 0.2 * \text{Lognormal}(x, 1, 1)\end{aligned} \tag{5.56}$$

The chosen Markov chain state transfer matrix, a uniform distribution spanning the range $[-5, 10]$, being n_1=1000 the minimum value for the number of state transfers and n_2=10000 the required number of samples. Subsequently, M-H sampling is

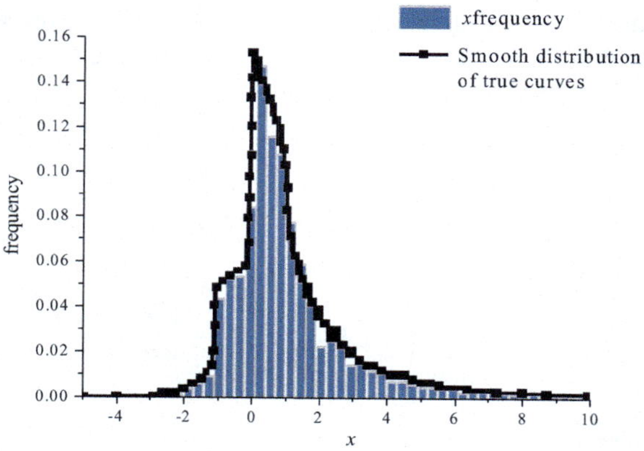

Fig. 5.10 Comparison of true curves for the sample set and the smooth distribution

conducted, and the resultant sample set is compared against the true curve of the smooth distribution, as illustrated in Fig. 5.10.

Upon examining the relationship between the distribution of the sampled values and the actual distribution, it becomes evident that the sampled values remain in good agreement with the corresponding stationary distribution. This observation confirms the suitability of the method for meeting the requirements of engineering applications.

5.2.3 *Unified Analysis and Propagation Method for Mixed Uncertainty Based on the aPC Method*

Building upon Sects. 5.2.1 and 5.2.2, this section outlines a unified analysis and propagation method for handling mixed uncertainty arising from stochastic and cognitive influences, leveraging the PCE method.

1. Establishment of PCE model under mixed uncertainty

To establish the PCE model under mixed uncertainty, the interval variable Z is incorporated into the coefficient of the PCE model, with the inclusion of the coefficient c_j. This is expressed as a quadratic polynomial of interval variables:

$$c_j = A_{j0} + \sum_{i=1}^{\lambda} B_{ji}\eta_i + \sum_{i=1}^{\lambda} D_{ji}\eta_i^2 \tag{5.57}$$

where λ is the number of interval variables; η_i is the normalisation of the interval variable z_i, $\eta_i = (z_i - z_i^c)/z_i^r$, with z_i^c and z_i^r as the midpoint and radius of the interval, respectively; A_{j0}, B_{ji}, and D_{ji} are coefficients determined by the sampling method.

For the strong statistical variable X and the sparse variable Y, orthogonal polynomials are embedded. First, both are transformed into standard uncertainties $\overline{X}$

and $\overline{Y}$. If the distribution of x_i follows ϑ and the distribution parameters are ξ_1 and ξ_2,then $\overline{x}_i=(x_i-\xi_1)/f\xi_2$; similarly, if the expression for y_i is $y_i=\sum_j w_{i,j}y_{i,j}=\sum_j w_{i,j}\vartheta_{i,j}(\xi_{1,ij},\xi_{2,ij})$, then $\overline{y}_i=\sum_j w_{i,j}(y_{i,j}-\xi_{1,ij})/\xi_{2,ij}$.

The PCE model considering 3 uncertainties can be expressed as:

$$Z(u)=Z(\overline{x}_1,\overline{x}_2,\cdots,\overline{x}_m;\overline{y}_1,\overline{y}_2,\cdots,\overline{y}_n;\overline{z}_1,\overline{z}_2,\cdots,\overline{z}_\lambda)\approx\sum_{i=0}^{M}c_i\Psi_i(u)$$
$$=\sum_{j=0}^{M-1}\left(A_{j0}+\sum_{i=1}^{\lambda}B_{ji}\eta_i+\sum_{i=1}^{\lambda}D_{ji}\eta_i^2\right)\Psi_j(u) \tag{5.58}$$

where $u=(\overline{x}_1,\overline{x}_2,\cdots,\overline{x}_m;\overline{y}_1,\overline{y}_2,\cdots,\overline{y}_n;\overline{z}_1,\overline{z}_2,\cdots,\overline{z}_\lambda)$ is the set of variables, the variables in u are independent of each other, m and n represent the number of variables for X and Y, respectively; $\Psi_i(u)$ is a multidimensional orthogonal polynomial basis; and the number of coefficients M depends on the number of variables $m+n$ and the expansion of order d, given by:

$$M=(m+n+d)!/((m+n)!d!) \tag{5.59}$$

2. PCE model coefficient solving

In resolving the coefficients of the PCE model, the LHS method is employed to sample the interval variable Z, generating sample points $z_1^w,z_2^w,\cdots,z_\lambda^w)$, $w=1,2,\cdots,K$ where K represents the number of sample points.

For each sample of Z $(z_1^w,z_2^w,\cdots,z_\lambda^w)$, the corresponding normalised value η^w is initially derived. Subsequently, the higher-order moments of X and Y are obtained via their PDFs. Referring to Eq. (5.34), the orthogonal polynomial basis corresponding to both variables is determined, leading to the identification of $\Psi_j(u)$. Finally, employing the regression method, the coefficient matrix c_j under the sample is obtained.

For c_j, it essentially constitutes a quadratic polynomial concerning η. Therefore, based on Eq. (5.57), the coefficients can be redefined as:

$$HE_j=c_j, j=1,2,\cdots,M \tag{5.60}$$

$$H=\begin{bmatrix}1 & (\eta_1)^1 & \cdots & (\eta_\lambda)^1 & (\eta_1^2)^1 & \cdots & (\eta_\lambda^2)^1\\ 1 & (\eta_1)^2 & \cdots & (\eta_\lambda)^2 & (\eta_1^2)^2 & \cdots & (\eta_1^2)^2\\ \vdots & \vdots & \ddots & \vdots & \vdots & \ddots & \vdots\\ 1 & (\eta_1)^K & \cdots & (\eta_\lambda)^K & (\eta_1^2)^K & \cdots & (\eta_1^2)^K\end{bmatrix} \tag{5.61}$$

$$E_j=\left[A_{j0}\ B_{j1}\ \cdots\ B_{j\lambda}\ D_{j1}\ \cdots\ D_{j\lambda}\right]^{\mathrm{T}} \tag{5.62}$$

$$c_j=\left[(c_j)^1\ (c_j)^2\ \cdots\ (c_j)^K\right]^{\mathrm{T}} \tag{5.63}$$

where H is the sample matrix determined by the known samples; and c_j is the coefficient matrix obtained via the regression method. Therefore, the coefficient matrix E_j can be derived using Eq. (5.60):

3. Unified analysis and propagation method for mixed uncertainty based on PCE method

The procedural flow of the unified analysis and propagation method for mixed uncertainty based on the PCE method is illustrated in Fig. 5.11.

Step 1: categorise the uncertainty quantities into strong statistical variables X, sparse variables Y, and interval variables Z based on the known sample data.

Step 2: sample the outer layer using LHS for Z. The sample points are $(z_1^w, z_2^w, \cdots, z_\lambda^w)$, where w=1, 2, $\cdots$, K represents the number of interval variables Z and K denotes the number of sample points.

Step 3: At each w sample point $(z_1^w, z_2^w, \cdots, z_\lambda^w)$ of the interval variable Z, based on the PDF of X and Y, the MC method and IPCM are utilised to sample X, whereas the M-H method is employed to sample Y. The sample points are $(x_1^\eta, x_2^\eta, \cdots, x_m^\eta, y_1^\eta, y_2^\eta, \cdots, y_n^\eta)$, η=1, 2, $\cdots$, N, where m denotes the number of strong statistical variables X, n denotes the number of sparse variables Y, and N denotes the total number of sample points for X and Y. Subsequently, the response S at this point and the corresponding orthogonal polynomial matrix Γ are computed. Utilising the regression method, the coefficient matrix CC^w at this stage is calculated as follows:

$$\Gamma=\begin{bmatrix} \Psi_0(\xi^{(1)}) & \Psi_1(\xi^{(1)}) & \cdots & \Psi_{M-1}(\xi^{(1)}) \\ \Psi_0(\xi^{(2)}) & \Psi_1(\xi^{(2)}) & \cdots & \Psi_{M-1}(\xi^{(2)}) \\ \vdots & \vdots & \vdots & \vdots \\ \Psi_0(\xi^{(N)}) & \Psi_1(\xi^{(N)}) & \cdots & \Psi_{M-1}(\xi^{(N)}) \end{bmatrix} \tag{5.64}$$

$$CC^w=(\Gamma^{\mathrm{T}}\Gamma)^{-1}\Gamma^{\mathrm{T}}S \tag{5.65}$$

Step 4: repeat Step 3 for each sample point of the interval variable Z until all sample points have been processed.

Step 5: construct the sample matrix H and coefficient matrix ($j = 0, 1, 2, \cdots, M-1$). For CC^w, the following expression applies:

$$\begin{aligned} CC^1 &= [C_0\, C_1\, C_2 \cdots C_{M-1}]^1 \\ CC^2 &= [C_0\, C_1\, C_2 \cdots C_{M-1}]^2 \\ &\vdots \\ CC^K &= [C_0\, C_1\, C_2 \cdots C_{M-1}]^K \end{aligned} \tag{5.66}$$

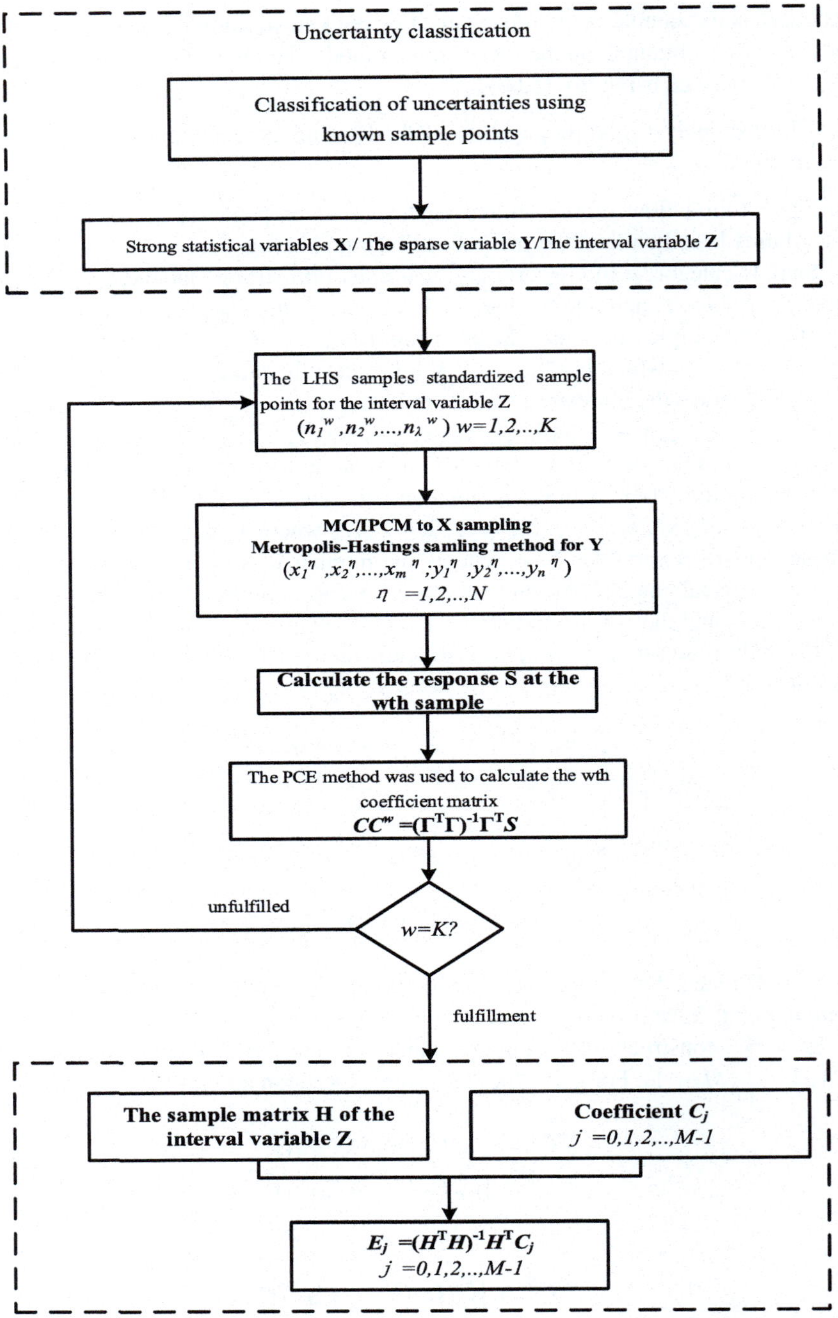

Fig. 5.11 The flow of the unified analysis and propagation method for mixed uncertainty based on PCE method

Each column of formula is drawn out, i.e., the same coefficients are taken under K different samples to construct the coefficient matrix c_j:

$$\begin{aligned} c_0 &= [(C_0)^1\ (C_0)^2\ \cdots\ (C_0)^K]^{\mathrm{T}} \\ c_1 &= [(C_1)^1\ (C_1)^2\ \cdots\ (C_1)^K]^{\mathrm{T}} \\ &\vdots \\ c_{M-1} &= [(C_{M-1})^1\ (C_{M-1})^2\ \cdots\ (C_{M-1})^K]^{\mathrm{T}} \end{aligned} \tag{5.67}$$

Step 6: solve Eq. (5.68) to derive the coefficient matrix E_j:

$$E_j = (H^{\mathrm{T}}H)^{-1}H^{\mathrm{T}}c_j \tag{5.68}$$

5.2.4 Numerical Example

The mathematical functions outlined in Sect. 3.2.3 are selected to validate the unified analysis and propagation method for mixed uncertainty based on aPC expansion method proposed in this chapter.

1. Consider only strong statistical variables, sparse variables

Consider the two-dimensional third-order nonlinear function of Eq. (3.17) with x_1 and x_2 independent. Unlike in Sect. 3.2.3.1, let x_1 be a strong statistical variable conforming to a normal distribution with a mean of 10 and standard deviation of 5. Let x_2 be a sparse variable with a PDF given by:

$$f(x_2) = w_1 f_{uniform} + w_2 f_{normal} = 0.001 f_{uniform} + 0.999 f_{normal} \tag{5.69}$$

where $f_{uniform} = 1/2; f_{normal} = \frac{1}{\sqrt{2\pi}\cdot 5} e^{-\frac{(x-9.9)^2}{2\cdot 5^2}}; -\infty < x < +\infty$.

5.2.3 The unified analysis and propagation method is applied to solve the PCE model in this scenario. As the uniformly distributed weight of x_2 is significantly smaller than that of the normally distributed weight, the PCE model coefficients established by the unified analysis and propagation method should closely approximate the coefficients of the real function in Sect. 3.2.3.1. The comparison between the coefficients of the calculated PCE model and of the real function is presented in Table 5.6.

The calculated coefficients demonstrate close proximity to the exact solution, thus initially verifying the accuracy of the method. Subsequently, the established PCE model's accuracy is tested. A set of test samples (10^3 samples) is chosen to forecast the model's response, and the results are juxtaposed with those of the real function, as depicted in Fig. 5.12. Comparing with the exact solution, the PCE model's response forecast for the test samples exhibits high accuracy, with almost all samples

Table 5.6 Comparison of model coefficients under the harmonised framework

Ratio	b_0	b_1	b_2	b_3	b_4	b_5	b_6	b_7	b_8	b_9
Exact solution	4682.3	2470.2	742.5	125	2865	500	0	997.5	125	125
Calculated value	4651.7	2442.7	732.9	124	2860	498	0	996.3	125	125

overlapping with the exact solution. This further validates the method's accuracy in handling mixed uncertainty through the established PCE model.

2. Consideration of strong statistical variables, sparse variables, and interval variables

Let us examine the Ishigami function defined by Eq. (3.20), where X_1 represents a strong statistical variable uniformly distributed on $[-\pi, \pi]$, X_3 represents an interval variable with upper and lower bounds of $-\pi$ and π, and X_2 represents a sparse variable with a PDF given by:

$$f(X_2) = w_1 f_{uniform} + w_2 f_{normal} = 0.001 f_{uniform} + 0.999 f_{normal} \tag{5.70}$$

where $f_{uniform} = 1/2\pi; f_{normal} = \frac{1}{\sqrt{2\pi}} e^{-\frac{x^2}{2}}; -\infty < x < +\infty$.

The PCE model is solved using the unified analysis and propagation method from Sect. 5.2.3. Due to the Ishigami function's strong nonlinearity, PCE models of orders 3, 5, 7, and 9 were constructed and compared with the results of the MC method, as illustrated in Table 5.7 and Fig. 5.13. Values in parentheses denote the percentage error of the mean and standard deviation concerning the MC method at different orders of expansion.

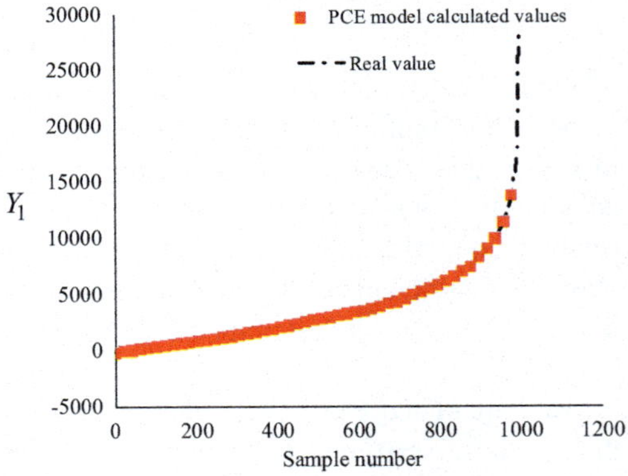

Fig. 5.12 PCE model accuracy validation

Table 5.7 Means and standard deviations of PCE model with different expansion orders

Expansion order	Average value	Standard deviation
$d = 3$	3.401 (0.3%)	2.719 (25.7%)
$d = 5$	3.397 (0.4%)	3.449 (5.8%)
$d = 7$	3.420 (0.2%)	3.631 (0.7%)
$d = 9$	3.419 (0.2%)	3.641 (0.5%)
MC	3.411	3.660

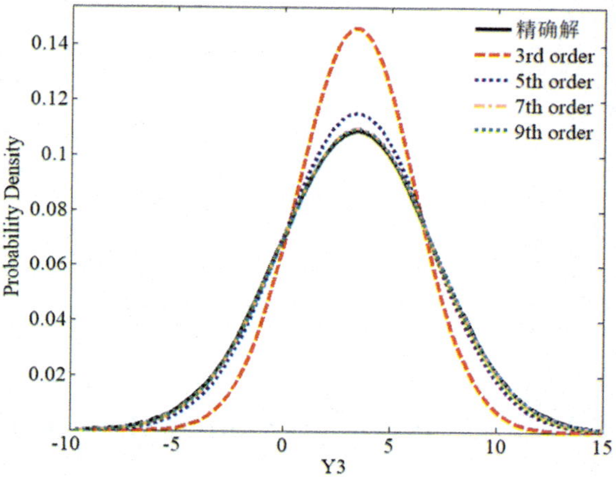

Fig. 5.13 Response PDF curves of PCE model with different unfolding orders

Table 5.7 illustrates that the mean errors of the PCE model at various expansion orders are all within the 1% range. Regarding standard deviation, lower expansion orders ($d = 3, 5$) yield larger prediction errors, while higher orders lead to convergence, with errors for 7th-order polynomials falling below 1%

A similar trend is observed in the PCE model response PDF curves depicted in Fig. 5.13. The peaks of the PDF curves align closely, with maximum probability density centred at $Y_3 = 3 \sim 4$. At order $d = 3$, the PCE model PDF curve exhibit larger errors; at $d = 5$, the low-probability region either end of the curve gradually approaches the exact solution curve; at $d = 7, 9$, both PCE models accurately quantify uncertainty of Y_3, with their PDF curves closely matching the exact MC solution curve, thereby confirming the accuracy of the PCE model established by the unified analysis and propagation method when considering the three types of uncertainty.

To further assess the established PCE model's accuracy, a set of test samples (10^3 in total) is used to forecast the model's response, and the results are compared with the real function outcomes, as shown in Fig. 5.14. Lower-order ($d = 3, 5$) PCE models exhibit larger errors in forecasting test sample responses compared to the exact solution. However, higher-order models ($d = 7, 9$) accurately forecast responses for most test samples, with almost all samples aligning closely with the

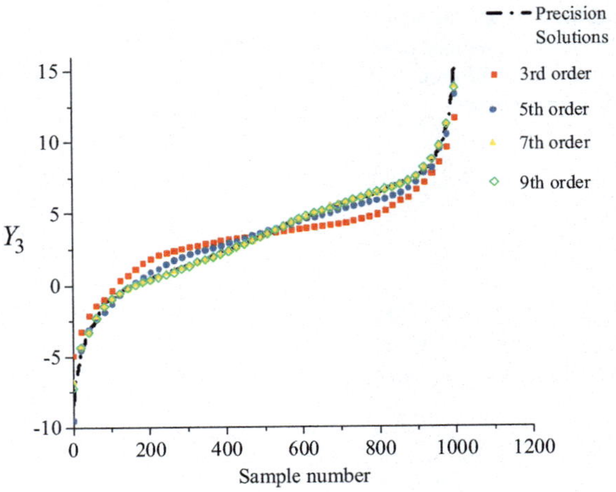

Fig. 5.14 PCE model accuracy validation

exact solution curve. This further validates the method's effectiveness in constructing PCE models to address the three uncertainties.

5.3 Ship Uncertainty Optimisation Design Example

To validate the viability of the two mixed uncertainty unified analysis and propagation methods introduced in this chapter, we have selected the bulk carrier optimisation design example from Sect. 3.5.1 of Chap. 3 and the KCS profile optimisation example from Sect. 3.5.2 for comparison. This comparison aims to elucidate the similarities and differences between the two methods.

5.3.1 Example of an Optimised Design for a Bulk Carrier

To commence the optimised design of bulk carriers, the three categories of uncertain parameters are initially reclassified, as illustrated in Tables 5.8 and 5.9.

For the interval variable representing the ship breadth exponent, two scenarios were identified to depict their BPA values, as outlined in Table 5.10. In Scenario 1,

Table 5.8 Uncertain parameters under aleatory uncertainty

	Speed fluctuation Z	Ship length exponent X	Ship width exponent Y
Variable type	Strong statistical variable	Strong statistical variable	Strong statistical variable
Distribution	Normal distribution	Normal distribution	Normal distribution
Parameter	μ=0,σ=0.1V_k	μ=1.7,σ=0.1μ	μ=0.7,σ=0.1μ

Table 5.9 Uncertainty parameters under mixed stochastic and cognitive uncertainty

	Speed fluctuations Z	Ship length exponent X	Ship width exponent YY
Variable type	Sparse variable	Strong statistical variable	Interval variable
Distribution	$0.01f_{uniform} + 0.99f_{normal}$	Normal distribution	[0.5,0.9]
Parameter	$U(0,1), N(0, 0.1V_k)$	μ=1.7,σ=0.1μ	BPA values are shown in the Table 5.10

Table 5.10 BPA values for the ship's breadth exponent

Case	Interval	BPA value	Number of points
1	[0.5, 0.9]	1	$10^2 \times 10^3$
2	[0.5, 0.66]	0.3	
	[0.66, 0.74]	0.4	$3 \times 10^2 \times 10^3$
	[0.74, 0.9]	0.3	

the PCE method was utilised, while both Scenarios 1 and 2 were analysed using the theory of evidence.

The design variables and constraints chosen in this chapter remain consistent with Sect. 3.5. However, for these two mixed uncertainty analysis and propagation methods, only the expressions for the objective function and reliability constraints differ, as indicated in Table 5.11, where:

$$TC_{Middle} = \frac{TC^L + TC^U}{2} \tag{5.71}$$

$$TC_{Radius} = \frac{TC^U - TC^L}{2} \tag{5.72}$$

Subsequently, the two mixed uncertainty unified analysis and propagation methods proposed in this chapter are employed to conduct RBRDO for this bulk

Table 5.11 Expression of objective function and reliability constraints under the influence of mixed uncertainty

Enterprise	PCE method	Evidence theory
Objective function	$\min \mu(TC)/\sigma(TC)$	$\min TC_{Middle}/TC_{Radius}$
Reliability constraints	$p_f(F_n \le 0.32) \le 1\%$	$Bel(F_n \le 0.32) \le 1\%$ $Pl(F_n \le 0.32) \le 1\%$
	$p_f(25000 \le DW \le 500000) \le 1\%$	$Bel(25000 \le DW \le 500000) \le 1\%$ $Pl(25000 \le DW \le 500000) \le 1\%$
	$p_f(T - 0.45DW^{0.31} \le 0) \le 1\%$	$Bel(T - 0.45DW^{0.31} \le 0) \le 1\%$ $Pl(T - 0.45DW^{0.31} \le 0) \le 1\%$

Table 5.12 Comparison of RBRDO results

	PCE method	Evidence theory
TC^L	–	6.036
TC^U	–	14.218
$TC/(\text{Pounds/yr})$	8.645	8.621
$\mu(TC)\ /TC_{Middle}/(\text{Pounds/yr})$	10.092	10.127
$\sigma(TC)\ /TC_{Radius}$	4.086	4.091
L/m	194.060	195.049
B/m	31.379	32.270
D/m	16.245	14.316
T/m	11.554	10.704
C_B	0.745	0.750
V_k/kn	14.107	14.020
$p_f(F_n \leq 0.32)$	0.1%	See Table 5.13
$p_f(25000 \leq DW \leq 500000)$	0.1%	See Table 5.13
$p_f(T - 0.45DW^{0.31} \leq 0)$	0.1%	See Table 5.13

carrier optimal design model. The optimal solutions obtained from the two methods are compared in Table 5.12, while the constrained failure probabilities of the optimal solutions, considering the uncertainty parameters, are presented in Table 5.13.

As evident from Tables 5.12, the optimal solutions derived from both methods of solving are relatively similar. However, the number of calculation points required by the theory of evidence far exceeds that needed by the PCE method, resulting in higher computational costs.

Moreover, owing to the presence of cognitive uncertainty, the ultimate form of the goal expressed based on the mixed uncertainty of the evidence theory is not a single exact value but is presented as an interval [6.036, 14.218]. Subsequently, through the midpoint and radius of the interval, the goal is expressed in terms of mean and standard deviation, as depicted in Table 5.12. Notably, the mean and standard deviation of the goal obtained by the evidence theory and the PCE method of solving are very close.

Tables 5.12 and 5.13 illustrate that the failure probability obtained by both methods is less than 1%, meeting the reliability requirements. The constraint failure probability derived from solving based on evidence theory is expressed by the interval [*Bel*,

Table 5.13 Constrained belief and plausibility of truth for evidence theory solutions

Austerity	Case 1		Case 2	
	Bel	*Pl*	*Bel*	*Pl*
$p_f(F_n \leq 0.32)$	0	0	0	0
$p_f(25000 \leq DW \leq 500000)$	0.00084	0.00116	0.00096	0.00104
$p_f(T - 0.45DW^{0.31} \leq 0)$	0.00088	0.00112	0.00093	0.00109

Pl], comprising belief *Bel* and plausibility *Pl*, both yielding a failure probability of less than 1%. It is noteworthy that the constraint failure probabilities computed by the PCE method consistently fall within the interval [*Bel*, *Pl*] for both Scenario 1 and Scenario 2. Notably, for the ship's breadth exponent Y, the greater number of subintervals and BPA values for Scheme 2 results in a failure probability interval [*Bel*, *Pl*] significantly smaller than that of Scheme 1. This indicates that obtaining more information about cognitive uncertainty leads to more accurate results. It is foreseeable that with continued increase in the number of subintervals and BPA values of the ship width exponent Y, and the portrayal of cognitive uncertainty in greater detail, the interval [*Bel*, *Pl*] will further shrink until converging to the exact probability value.

To observe the change in failure probability more intuitively, the constraints of *c* under different limit values are calculated. Subsequently, the CDF (PCE method), cumulative belief distribution function, and cumulative plausibility distribution function (evidence theory) curves of the constraints of the optimal scheme obtained by the two methods are obtained, denoted as $1 - p_f$ (i.e., the probability of success), as shown in Figs. 5.15, 5.16, 5.17. It is apparent that the three constrained failure probabilities obtained by the two methods satisfy the reliability requirements at the given limits ($c_{F_n} = 0.32, c_{25000-DW} = 0, c_{T-0.45DW^{0.31}} = 0$). Due to the presence of strong statistical variables and sparse variables, the curves of *Bel* and *Pl* are continuous instead of stepwise when only interval variables are considered. Additionally, the reliability exponents no longer change abruptly with *c*; instead, they change continuously. Furthermore, for these three constraints, both curves *Bel* and *Pl* for Scheme 1 enclose both curves *Bel* and *Pl* of Scheme 2, reaffirming the conclusion that greater information about cognitive uncertainty leads to more accurate results. Moreover, compared to the previous numerical case in 4.2.3, the spacing between the curves *Bel* and *Pl* is smaller in this case, indicating that cognitive uncertainty has a lesser impact on the reliability calculation results.

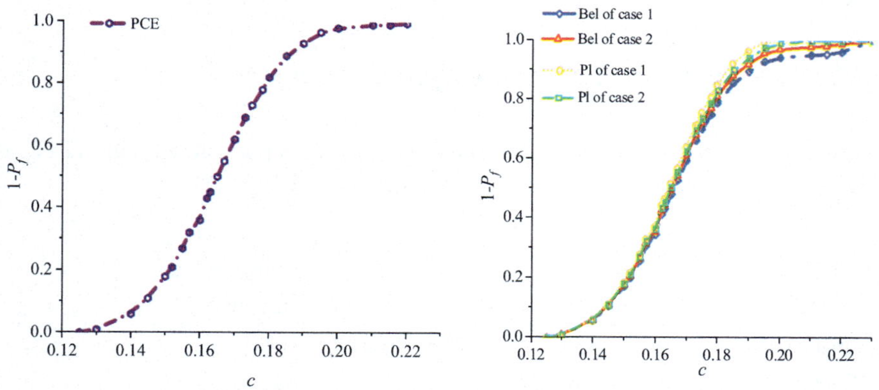

(a) Cumulative dens ity (b) Belief and plausibility

Fig. 5.15 $F_n \leq c$ Cumulative density **a**, belief and plausibility curves **b**

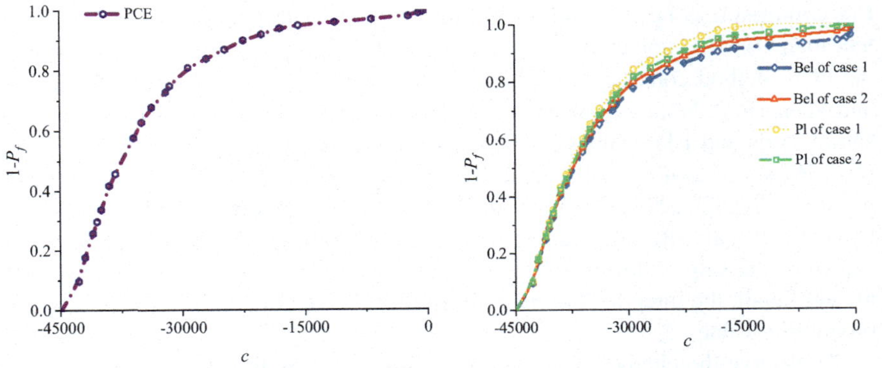

(a) Cumulative density (b) Belief and plausibility

Fig. 5.16 $25000 - DW \le c$ Cumulative density **a**, belief and plausibility curves **b**

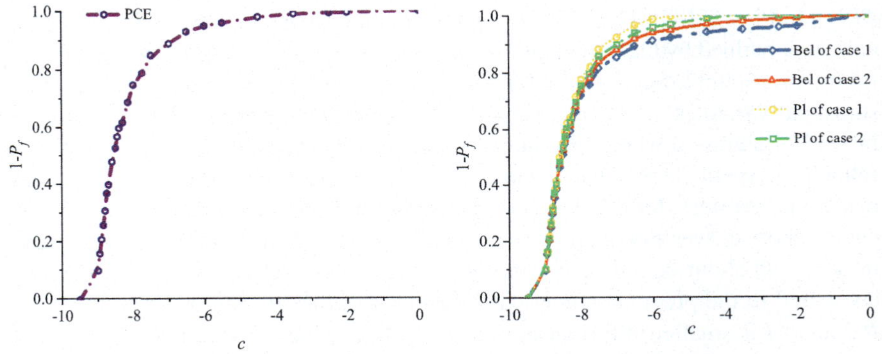

(a) Cumulative density (b) Belief and plausibility

Fig. 5.17 $T - 0.45DW^{0.31} \le c$ Cumulative density **a**, belief and plausibility curves **b**

5.3.2 Example of KCS Profile Optimisation

In this section, the KCS, one of the standard ship types published by the ITTC, is chosen as the subject of study, referencing Sect. 3.5.2 Table 5.14.

Table 5.14 Uncertainty parameters for mixed stochastic and cognitive uncertainty

	Speed fluctuation v	Draught fluctuation t
Variable type	Sparse variable	Interval variable
Distribution	$0.01f_{uniform} + 0.99f_{normal}$	[0.336, 0.347]
Parameter	$U(0, 1), N(0, 0.1V_k)$	BPA values are shown in the Table 5.15

Table 5.15 BPA values for draught fluctuations

Case	Interval	BPA value	Number of points
1	[0.336, 0.347]	1	$10^2 \times 10^3$
2	[0.336, 0.340]	0.2	
	[0.340, 0.343]	0.6	$3 \times 10^2 \times 10^3$
	[0.343, 0.347]	0.2	

Similar to the process employed for the optimal design of bulk carriers, the uncertain parameters, represented by speed fluctuations and draft fluctuations, are initially redefined:

Likewise, for the interval variable draught fluctuation t, two scenarios are identified to express their BPA values, detailed in Table 5.15. Scenario 1 is utilised for the PCE method, while both scenarios 1 and 2 undergo calculations based on the theory of evidence.

The design variables and constraints selected for this example remain consistent with those outlined in Sect. 3.5.2. Only the objective function and reliability constraints, influenced by uncertainties, are altered, as demonstrated in Table 5.16.

Subsequently, the two mixed uncertainty unified analysis and propagation methods proposed in this chapter are applied to conduct RBRDO on the KCS model, with the optimisation results presented in Table 5.17. Simultaneously, the failure probabilities of the reliability constraints under varying EEDI thresholds are computed to obtain the CDF of the constraints (PCE method), cumulative belief distribution function, and CPF (Theory of Evidence) curves, illustrated in Fig. 5.18.

Both Table 5.17 and Fig. 5.18 illustrate that the failure probability of EEDI constraints is below 5%, meeting the reliability requirement. Concurrently, both curves *Bel* and *Pl* of Scheme 1 encompass those of Scheme 2, corroborating the

Table 5.16 Expression of objective function and reliability constraints under the influence of mixed uncertainty

	PCE method	Evidence theory
Objective function	$\min \mu(EEDI)/\sigma(EEDI)$	$\min EEDI_{Middle}/EEDI_{Radius}$
Reliability constraints	$P_f(EEDI < EEDI(I)) \le 5\%$	$Bel(EEDI < EEDI(I)) \le 5\%$ $Pl(EEDI < EEDI(I)) \le 5\%$

Table 5.17 Comparison of RBRDO results

	PCE method	Evidence theory
$\mu(EEDI)/EEDI_{Middle}$	17.02	17.11
$\sigma(EEDI)/EEDI_{Radius}$	4.32	4.41
$P_f(EEDI < EEDI(I))$	4.9%	case 1:$[Bel,Pl] = [0.001, 0.05]$ case 2:$[Bel,Pl] = [0.016, 0.035]$

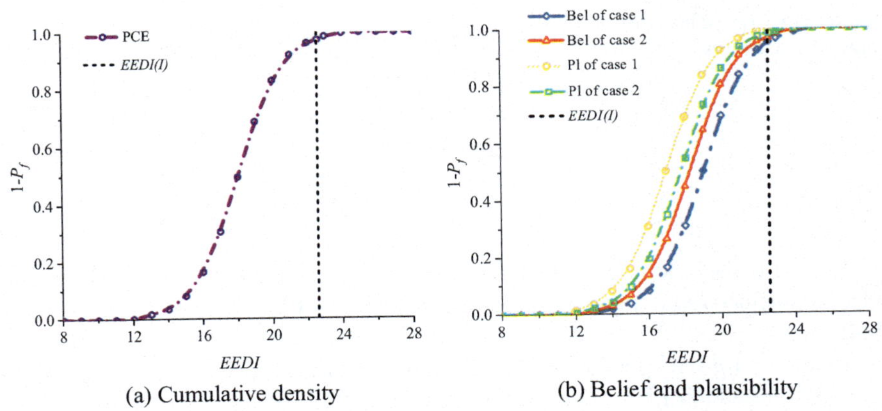

Fig. 5.18 EEDI Cumulative density **a**, belief and plausibility curves **b**

notion that increased cognitive uncertainty information leads to a narrower interval $[Bel, Pl]$ and more accurate failure probability.

It is noteworthy, as indicated in Table 5.17, that unlike the optimal design of bulk carriers, the constraint failure probability computed by the PCE method in this instance falls within the interval $[Bel, Pl]$ calculated by Scheme 1. However, the result obtained by Scheme 2 does not fall within the interval, rendering the conclusion invalid. This outcome can be attributed to computational errors arising during the optimisation analysis process, primarily stemming from a lack of relevant data for the uncertainty parameters in this scenario. Furthermore, for the interval variables of Scheme 2, parameter forms are empirically determined, making them susceptible to parameter form and BPA delineation errors. This underscores the significance of accurately classifying uncertainties before undertaking uncertainty optimisation.

References

1. Rajashekhar MR, Ellingwood BR. A new look at the response surface approach for reliability analysis[J]. Struct Saf. 1993;12(3):205–20.
2. Lindsay BG. On the determinants of moment matrices[J]. Ann Statist. 1989; 17(2) 711–21.
3. Karlin S. Total positivity[M]. Stanford: Stanford University Press; 1968. p. 576.

Chapter 6
Hull Form Optimisation Under Mixed Uncertainty

Hull form optimisation presents a quintessential uncertainty optimisation challenge. This chapter begins by synthesizing the findings from preceding chapters on uncertainty optimisation design. It then advances mathematical models for hull form uncertainty optimisation design and delineates a comprehensive process for undertaking such optimisation. Throughout this chapter, we maintain our focus on the KCS vessel, employing the mixed uncertainty analysis and propagation methodology introduced in Chap. 5. Our objective is to systematically investigate the influence of multiple uncertainties on the robustness and reliability of the KCS hull form. Diverging from Chap. 5, this chapter expands the scope by incorporating external environmental factors, specifically wave parameters, alongside ship speed and draught, into the uncertainty analysis. The uncertainty factors include wavelength and wave height, which are meticulously categorised through data collection to facilitate comprehensive uncertainty modelling. Our optimisation objective centres on the ship's mean resistance coefficient under top wave conditions, while the wave-resistance scale index serves as a reliability constraint. For the mixed uncertainty analysis and propagation, we employ both evidence theory and the PCE method. This amalgamation enables us to conduct a thorough hull form optimisation, taking into account the complexities arising from multiple sources of uncertainty.

6.1 Mathematical Models and General Process for Hull Form Optimisation with Uncertainty

Hull form uncertainty optimisation diverges from deterministic hull form optimisation concerning its objectives, constraints, and solution methodology. The inclusion of uncertainty analysis necessitates a distinct approach. Drawing from the insights gleaned in previous chapters, we propose general mathematical models and a comprehensive process for hull form uncertainty optimisation.

Z. Liu et al., *Uncertainty-Based Ship Design Optimization*,
https://doi.org/10.1007/978-981-97-2973-9_6

Table 6.1 Hull form uncertainty optimisation mathematical model

	RDO	RBDO	RBRDO
Design variable	X	X	X
Constraint	$g_i \leq 0$ $h_j = 0$	$P(g_i \leq 0) \geq r$ $\mu(h_j) = 0$	$P(g_i \leq 0) \geq r$ $\mu(h_j) = 0$
Objective function	$\mu(F)$ or/and $\sigma(F)$	F	$\mu(F)$ or/and $\sigma(F)$

6.1.1 Mathematical Models

Hull form uncertainty optimisation typically encompasses three main categories: RDO, RBDO, and RBRDO. The mathematical models for these optimisation approaches were previously outlined in Chap. 1 and are consolidated in Table 6.1.

6.1.2 General Process

The general solution process for ship uncertainty design optimisation is depicted in Fig. 6.1 and can be segmented into the following primary steps:

Step 1: Uncertainty system modelling. Uncertainty system modelling serves as preparatory work for uncertainty design optimisation, encompassing optimisation modelling and uncertainty factor analysis and modelling.

Step 1.1: Optimisation modelling. Optimisation modelling precedes design optimisation, involving the modelling of the research object system and its constituent disciplines, definition of the optimisation problem, and establishment of design variables, constraints, optimisation objectives, and system parameters. Within uncertainty system modelling, the optimisation objective includes a robustness optimisation objective, while the constraints encompass reliability constraints.

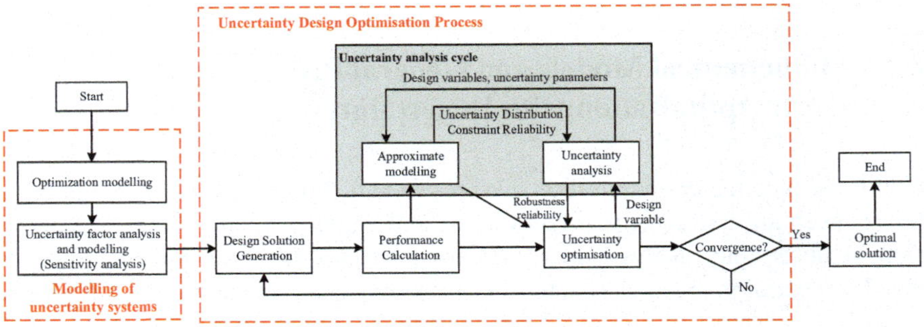

Fig. 6.1 Generic solution process for uncertainty design optimisation

Step 1.2: Uncertainty factor analysis and modelling. The initial stage involves analysing the uncertainties present in the optimisation problem and employing suitable mathematical methods to describe and quantify them. During this process, the numerous uncertainties involved are often scrutinised for sensitivity, with less impactful factors being eliminated to reduce the complexity of uncertainty design optimisation.

Step 2: Uncertainty design optimisation process. The uncertainty optimisation process comprises five steps: design solution generation, performance calculation, approximate modelling, uncertainty analysis, and uncertainty optimisation.

Step 2.1: Design solution generation. Design solutions are generated by adjusting the design variables. In the ship conceptual design stage, modifications to the main scale of the ship yield design solutions for subsequent stages. Meanwhile, in the profile design stage, adjustments to the hull's shape parameters generate different hull geometry models for further processing.

Step 2.2: Performance calculation. Empirical formulas or simulation calculation methods are utilised to compute the ship's performance indicators throughout the optimisation design process.

Step 2.3: Approximate modelling. Given the large number of sample calculations required in the uncertainty analysis process, some disciplines resort to high-precision calculations, necessitating a vast number of computations that may be prohibitively costly. Employing an approximate model offers an effective means of achieving rapid predictions to replace high-precision calculation methods. Additionally, for ship multidisciplinary design optimisation problems, approximation models are commonly employed to decouple disciplines or facilitate parallel computation.

Step 2.4: Uncertainty analysis. Uncertainty optimisation mandates quantitative analysis of the distribution characteristics of the system's performance under uncertainty influence for a given design solution. This analysis assesses the robustness and reliability of the design solution to meet constraints, necessitating input of design variables and uncertain parameters into the approximate model to swiftly predict design solution performance. Mature uncertainty analysis methods encompass stochastic uncertainty analysis transfer, reliability analysis, non-probabilistic uncertainty analysis, and hybrid uncertainty analysis methods. In this chapter, the hybrid uncertainty analysis method is employed for the KCS.

Step 2.5: Uncertainty optimisation. Diverging from deterministic optimisation, uncertainty analysis is nested within the search process to evaluate the robustness and reliability of each search point. This nested approach enables identification of optimal solutions within the design space based on robustness and reliability requirements. This section employs a two-layer loop algorithm to integrate uncertainty analysis directly into the optimisation, as delineated in Sect. 6.4.2.

Table 6.2 KCS model dimensions

L_{pp}/m	B_{wl}/m	D/m	T/m	C_b
6.0702	0.8498	0.5015	0.2850	0.651

6.2 CFD for KCS

6.2.1 KCS Model

The KCS model remains utilised in this chapter. In contrast to Sect. 3.5.2, the dimensions of the KCS model chosen here align with those selected for the wave resistance experiments. Its principal scales are detailed in Table 6.2.

6.2.2 CFD Settings and Results

1. **Turbulence modelling, free liquid surface, and wave modelling**

The commercial software STARCCM + was utilised for CFD calculations. The governing equations for incompressible turbulence consist of the continuity equation and the Reynolds-Averaged Navier–Stokes (RANS) equation, with the $SST\ k - \omega$ model chosen as the turbulence model to provide closure to the RANS equation. Widely applied in numerical ship hydrodynamics, this model is extensively used for simulating ship motions and resistance calculations.

The Volume of Fluid (VOF) method was selected to capture the free surface of the wave. This method, a straightforward multiphase flow model, resolves interfaces between multiple immiscible fluids [1]. The Semi-Implicit Method for Pressure-Linked Equations (SIMPLE) algorithm was chosen to couple the velocity field and pressure, serving as a common numerical method for solving the Navier–Stokes equations. It resolves the pressure correction equation iteratively until the continuity equation is satisfied [2].

In employing the RANS-VOF method, the Dynamic Fluid-Body Interaction (DFBI) module predicts the ship's motion response in waves. Through boundary condition updates, the ship can sway longitudinally and pendulously during calculations. The RANS solver computes the excitation forces and moments acting on the hull's surface at each time step and solves the ship's equations of motion to determine acceleration, velocity, and displacement [3]. The hull's position can be adjusted by updating the free surface position for grid shifting based on the hull's position and the velocity inlet's two-phase flow distribution [4].

The wave model selected for this study is a first-order Stokes wave, with its wavefront equation expressed as:

$$\zeta = \zeta_a \cos(kx - \omega_e t) \tag{6.1}$$

where ζ_a is the wave amplitude, k is the number of waves, ω_e is the frequency of the encounter; and k and ω_e are defined as:

$$k = \frac{2\pi}{\lambda} \tag{6.2}$$

$$\omega_e = \omega_0 + kU \tag{6.3}$$

where λ is the wavelength, U is the ship's speed, and ω_0 is the intrinsic frequency, defined as:

$$\omega_0 = \sqrt{2\pi g/\lambda} \tag{6.4}$$

To prevent echo effects at the boundary of the computational domain from influencing results, a damping area is set at the wave's exit boundary. Resistance is introduced in the vertical motion so that as the wave passes through this damping area, its height gradually diminishes until it becomes a static water surface. The equation governing its vertical direction is:

$$S_z^d = \rho(f_1 + f_2|w|)\frac{e^{\kappa} - 1}{e^1 - 1}w \tag{6.5}$$

where

$$\kappa = (\frac{x - x_{sd}}{x_{ed} - x_{sd}})^{n_d} \tag{6.6}$$

w is the vertical velocity; x_{sd} is the starting point of the wave cancellation area; x_{ed} is the end point of the wave cancellation area (usually the boundary); and f_1, f_2 and n_d are the wave cancellation parameters.

2. **Selection of the time step**

The Courant-Friedrichs-Lewy (CFL) number represents the ratio of the physical time step to the grid convective time scale, indicating the relationship between the grid cell size and the grid flow velocity. This can be expressed as:

$$CFL = U_m\frac{\Delta t}{\Delta x} \tag{6.7}$$

where U_m is the mesh flow rate, Δt is the time step, and Δx is the mesh cell size. A larger CFL value reduces numerical accuracy, whereas a smaller CFL value increases convergence time. Ideally, all locations within the computational domain should have a CFL value less than 1. It is important to note that different regions in the computational domain may have varying Δx and U_m values, leading to different CFL values.

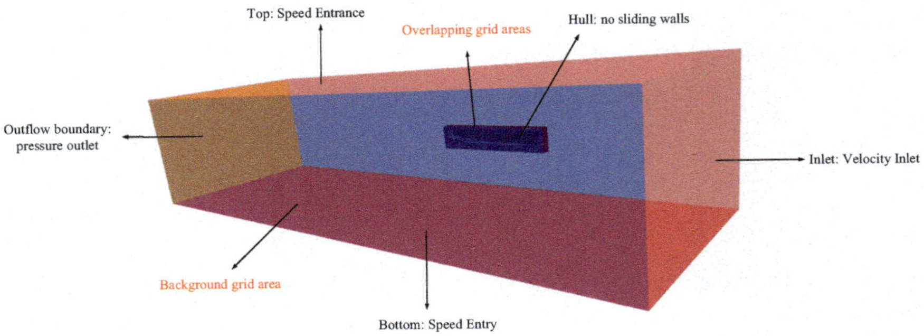

Fig. 6.2 Computational domain and boundary conditions

To accurately simulate ship motion response in regular waves, the ITTC [5] recommends using a minimum of 100 time steps per encounter cycle on the CFD model. CD-Adapco also suggests that the time step should be less than $1/80T \sim 1/100T$ when calculating first-order wave conditions. Therefore, this study adopts $\Delta t = 0.01$ s as the time step for simulating all cases, satisfying the requirement of at least 80–100 time steps in each encounter cycle.

3. **Computational domain and boundary conditions**

To enhance computational efficiency while studying ship performance in waves, the hull half-mode was selected for investigation, aiming to reduce the number of cells. Utilising overlapping grids to simulate ship motion entails the creation of two distinct regions: the background region and the overlapping region, illustrated in Fig. 6.2.

The choice of boundary conditions is pivotal in achieving accurate results without incurring unnecessary computational costs. While various boundary conditions are available, selecting the most suitable ones is imperative. Figure 6.2 delineates the boundary conditions of the computational domain.

Ensuring stable overlapping mesh computation necessitates effective information exchange between the background region and the overlapping region. STAR-CCM+ offers several interpolation methods, including distance-weighted, linear, and least-squares methods. After weighing computational accuracy and cost, the distance-weighted method is employed to interpolate data between the background mesh and overlapping mesh.

Figure 6.3 provides a side view perspective of the computational domain dimensions. According to ITTC guidelines for regular wave simulations, the inlet boundary should be situated at a distance of 1–2 L_{pp} from the hull, while the outlet boundary should be positioned 3–5 L_{pp} downstream to mitigate wave reflection effects. In this study, the inlet boundary of the computational domain is positioned 1 L_{pp} away from the hull, while the outlet boundary is located 7 L_{pp} from the hull, incorporating a 2 L_{pp} wave-absorbing zone. Additionally, the ship's side is positioned 3 L_{pp} from the side boundary of the computational domain, and the ship's bottom is situated 2 L_{pp} above the bottom of the computational domain.

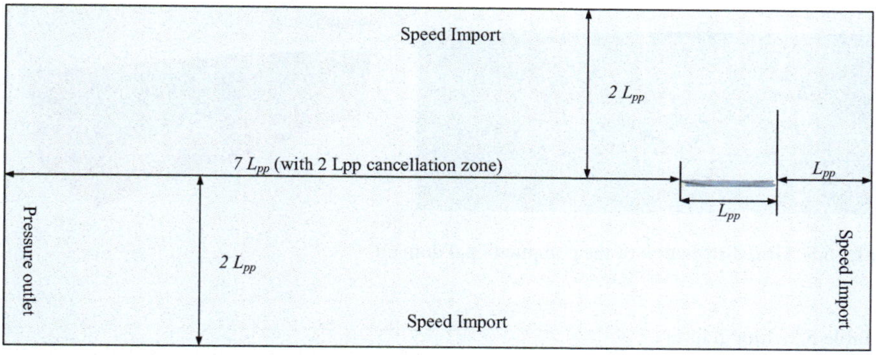

Fig. 6.3 Side view of the computational domain

4. **Grid generation**

Overlapping grids represent a novel grid generation approach applied in numerical predictions of ship motion responses in waves [6, 7]. Throughout the computational simulations in this chapter, overlapping grids are utilised for grid generation. The grid is segmented using a cut-body grid, which is denser around the hull and in the free-surface region. Moreover, the book adopts a grid with at least 100 cells per wavelength on the free surface and 10–20 cells in the vertical direction, resulting in a total computational grid of approximately 4.5 million grid cells.

The left panel of Fig. 6.4 illustrates the cross-section on the free surface surrounding the hull. The denser grid in proximity to the free surface ensures accurate capture of waveforms. Conversely, the right panel of Fig. 6.4 displays the grid distribution on the hull's surface, while Fig. 6.5 depicts the grid distribution across the computational domain.

5. **Comparison of CFD and experimental results**

In this book, initial numerical computations are conducted for the KCS scalar modes under top wave conditions, followed by a comparison of the results with experimental data. The specific operational parameters are outlined in Table 6.3.

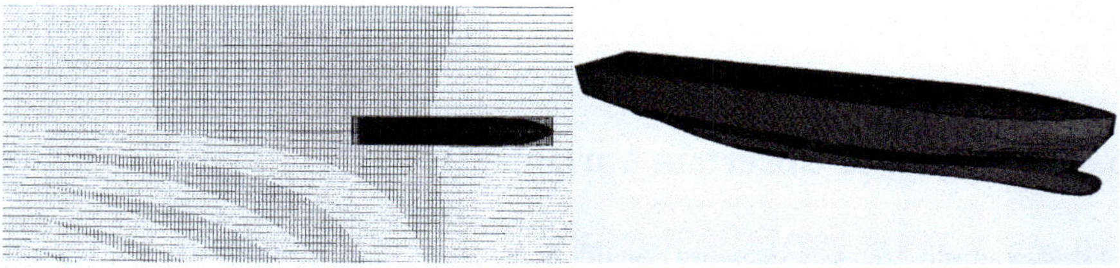

Fig. 6.4 Free liquid surface grid (left panel) and hull surface grid (right panel)

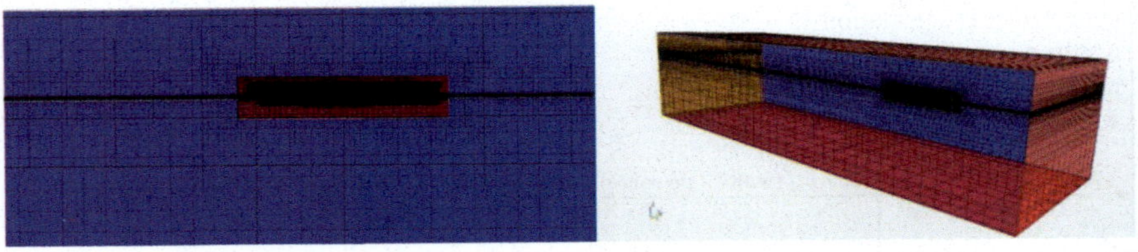

Fig. 6.5 Grid distribution of the computational domain

Table 6.3 Liquid level conditions for calculation working conditions

Case	λ/m	*H*/mm
Case 0(静水)	–	–
Case 1	3.949	62
Case 2	5.164	78
Case 3	8.321	149
Case 4	11.840	196

Table 6.4 Comparison between CFD calculation results and experimental results under different working conditions

Case	CFD	EFD	E%D (%)
Case 0(静水)	3.792	3.835	1.13
Case 1	8.553	8.253	−3.63
Case 2	9.060	9.244	1.99
Case 3	14.063	13.955	−0.77
Case 4	10.643	10.842	1.83

Numerical computations for the five conditions listed in Table 6.3 are executed with consistent grid division and time step configurations, while maintaining a Fourier number of $\mathrm{Fr} = 0.26$.

This chapter aims to utilise the average resistance of the ship under top wave conditions as the performance metric. Consequently, the average drag coefficient magnitude across the ship cycle is selected as the resultant computation upon convergence. The comparison between experimental and numerical results is presented in both Table 6.4 and Fig. 6.6.

6.3 Selection of Uncertain Variables

In this example, two aspects were considered: wave parameters and ship parameters, leading to the selection of four uncertain variables. Concerning wave parameters, the wave length λ and wave height (H) were chosen as uncertain variables. Regarding ship parameters, the ship speed (V) and draft (T) were selected as uncertain variables.

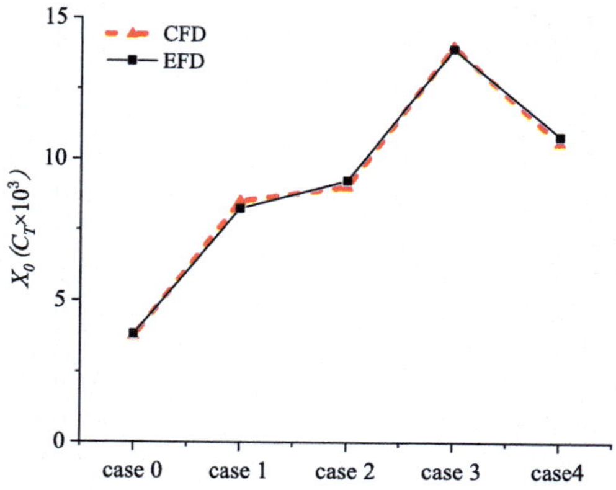

Fig. 6.6 Comparison between CFD calculation results and experimental results under different working conditions

6.3.1 Data of Uncertain Variables

For ship parameters, the navigation history of a specific ship for the past three days could be obtained through the AIS, providing details such as the current position, speed, and recording time. However, draft information was not included in the recorded data and was only available at the beginning of the voyage.

To enhance the data's applicability, this study observed data from five sister container ships sailing in the Mediterranean over a period of approximately two months. These ships followed very similar routes, with the only variation being that the third ship navigated in the southern Mediterranean. Notably, the tonnage and capacity of these five ships were nearly identical. Given the near-identical nature of the observed vessels, their data could be amalgamated to enhance the statistical robustness of the database.

To illustrate the ship's speed variations during a single voyage, Fig. 6.7 displays the ship speed fluctuations over time. It is evident from the figure that the ship's speed during navigation was far from constant, with maximum speed typically reached when approaching the port.

Upon obtaining the sample of selected ship routes, a statistical analysis of actual operating conditions was conducted. Each of the five sister ships provided approximately 10,000 data points, encompassing details such as the current speed, position, and direction of the ship, as outlined in Table 6.5.

Each dataset was segmented by position and speed to identify the commencement and conclusion of each voyage, thereby facilitating the identification of periods when a ship was docked in port. Additionally, for each voyage, the initial draft of the ship was recorded, as depicted in Table 6.6.

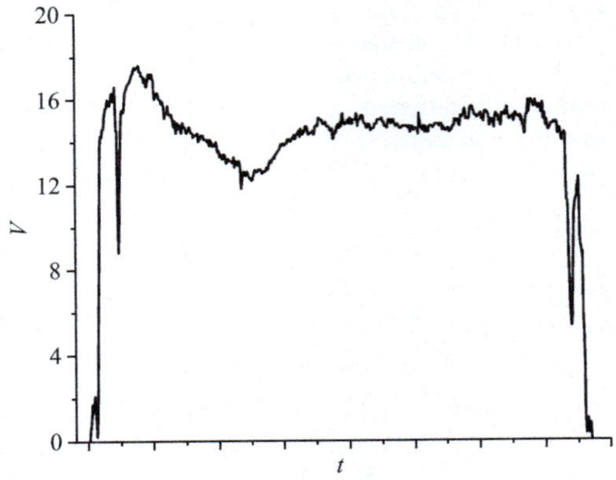

Fig. 6.7 Ship speed during a single voyage

6.3.2 *Modelling of Uncertainty*

Drawing upon the data presented in Tables 6.5 and 6.6, an uncertainty classification method was employed to ascertain the distribution of speed and draft parameters.

Utilising the speed data from the five sister ships, the CDF curve was plotted, as depicted in Fig. 6.8. This figure illustrated that the vessel remained anchored in port for approximately 10% of a single voyage cycle. Additionally, the percentage of ship speeds below 10 knots ranged between 15 and 20%, representing the proportion of the navigation cycle occupied by port operations.

To better capture the characteristics of speed during ship operations, this study exclusively considered data samples exceeding 10 knots. Employing the data, an uncertainty classification method was employed to determine the speed distribution. It was assumed that sailing speed adhered to six competing distributions of strong statistical variables and one mixed distribution of sparse variables, with distribution parameters estimated as detailed in Table 6.7. Subsequently, based on the CDF of these seven distributions, the Anderson–Darling test was conducted to calculate the OSL value for each distribution, as outlined in Table 6.6. Notably, the distribution corresponding to the maximum OSL value for speed was identified as the normal distribution, thus, determining the speed to be a strong statistical variable following $N(15.555, 1.616)$.

Similarly, for draft, the OSL values for the seven distributions are detailed in Table 6.8. It is observed that the maximum OSL value corresponds to a mixed distribution, signifying draft as a sparse variable. The parameters and weights of its mixed distribution are presented in Table 6.9.

Given that the aforementioned data pertained to the actual speed and draft of the vessel, it was imperative to convert these data into the corresponding speed and draft of the ship model (Table 6.10). The specific draft format is delineated in Table 6.11.

Regarding wave parameters, such as wave length and wave height spectrum, selection was based primarily on experience due to the paucity of adequate data. Typically, the wave length was chosen within the range of $0.75L_{pp} \sim 2L_{pp}$, rendering

Table 6.5 Speed data of 5 sister ships

Ship1					
Number	Year-month-day-hour-minute	Speed(km)	Latitude	Longitude	Course
1	2014 12 22 09 01	18	36.64796	21.27539	287°
2	2014 12 22 08 58	17.8	36.64307	21.29484	288°
…	…	…	…	…	…
…	…	…	…	…	…
10,288	2014 11 19 12 51	0	40.67093	14.74059	216°
10,289	2014 11 19 12 27	0	40.67095	14.74065	216°
Ship2					
1	2014 12 22 09 08	3.9	53.62962	6.14161	257°
2	2014 12 22 09 05	3.6	53.63052	6.146598	248°
…	…	…	…	…	…
…	…	…	…	…	…
10,114	2014 11 19 10 16	13.6	51.31091	-3.924792	81°
10,115	2014 11 19 10 13	13.5	51.30917	-3.942523	81°
Ship3					
1	2014 12 22 05 46	15.7	35.99749	-6.352125	284°
2	2014 12 22 05 43	15.6	35.99397	-6.336525	285°
…	…	…	…	…	…
…	…	…	…	…	…
6715	2014 11 19 2 4	0	50.89979	-1.41516	289°
6716	2014 11 19 1 43	0	50.89972	-1.41516	289°
Ship4					
1	2014 12 22 09 01	0	41.332	2.146333	317°
2	2014 12 22 08 37	0	41.332	2.146333	294°
…	…	…	…	…	…
…	…	…	…	…	…
8871	2014 11 19 10 25	16.4	50.3105	-1.780333	63°
8872	2014 11 19 10 22	16.5	50.30433	-1.7995	63°
Ship5					
1	2014 12 22 09 21	13.5	51.44595	-4.42867	94°
2	2014 12 22 09 19	13.3	51.44653	-4.441578	95°
…	…	…	…	…	…
…	…	…	…	…	…
9268	2014 11 19 10 40	16	36.03896	-4.883	81°
9269	2014 11 19 10 38	16	36.03741	-4.896564	81°

Table 6.6 Draft data of 5 sister ships

	Ship1	Ship2	Ship3	Ship4	Ship5
1	7.7 m	8.7 m	8.4 m	8.3 m	8.3 m
2	7.9 m	8.5 m	8.3 m	7.8 m	7.7 m
…	…	…	…	…	…
…	…	…	…	…	…
22	8.4 m	8.1 m	7.7 m	8.4 m	7.0 m
23	8.3 m	8.4 m	7.7 m	7.4 m	8.2 m

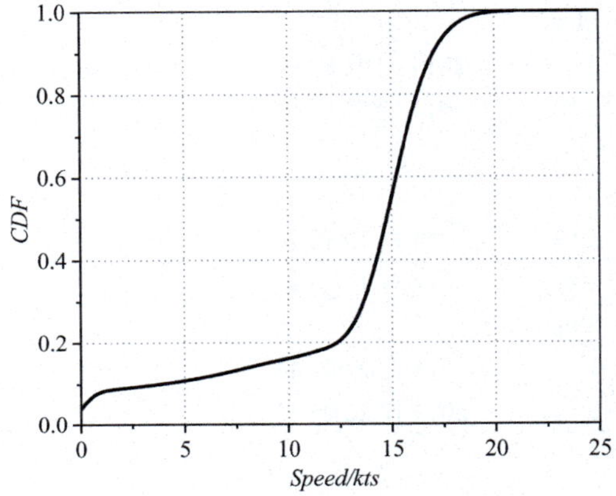

Fig. 6.8 CDF curve of the speed

Table 6.7 Speed V /kn

Distribution	ξ_1	ξ_2	OSL
Normal	**15.555**	**1.616**	**0.847**
Uniform	10.000	25.600	0
Exponential	15.555	0.064	0
Weibull	16.272	10.115	0
Extreme	14.729	1.783	0
Lognormal	2.739	0.107	0.665
Mixed	–	–	0

Table 6.8 OSL values for draught

Distribution	Normal	Uniform	Exponentia	Weibull	Extreme	Lognormal	Mixed
OSL	0.005	0.000	0.000	0.058	0.000	0.001	**0.141**

Table 6.9 Mixed distribution of the draft T

Distribution	ξ_1	ξ_2	Weight
Normal	8.013	0.444	0.413
Uniform	7.000	9.000	0.000
Exponential	8.013	0.125	0.000
Weibull	8.218	20.655	0.441
Extreme	7.786	0.443	0.000
Lognormal	2.080	0.056	0.146

Table 6.10 Uncertainties in seakeeping calculations

Uncertain variables	Type	Distribution
V(m/s)	Strong statistic variable	Normal $N(2.00, 0.21)$
T(m)	Sparse variable	Shown in Table 3.8
λ(m)	Interval variable	[4.55, 12.14]
H(m)	Interval variable	[0.03, 0.10]

Table 6.11 Expression of draft T

Distribution	ξ_1	ξ_2	Weight
Normal	0.285	0.016	0.744
Uniform	0.249	0.320	0
Exponential	0.285	3.509	0
Weibull	0.292	20.708	0.152
Extreme	0.277	0.016	0
Lognomal	−1.257	0.056	0.104

the wave length an interval variable of [4.55, 12.14]. As for wave height, it generally equated to 1/10 of the depth D, with a maximum not exceeding 1/5. Hence, the interval for wave height was determined to be [0.03, 0.10].

In conclusion, the inclusion of the four selected uncertain variables—speed, draft, wave length, and wave height—incorporating three different forms of uncertainty, contributed to enriching the final KCS hull form optimisation with mixed uncertainties.

6.4 Optimisation Definition

6.4.1 Objective Function, Design Variables, and Constraints

In this chapter, the KCS serves as the subject for RBRDO concerning resistance in waves. The mean resistance coefficient $\overline{C}_T$ of the ship under top wave conditions was chosen as the performance objective, with both the expected value and standard deviation $\overline{C}_T$ of the coefficient selected for optimisation. The design variables remain consistent with those outlined in Sect. 3.5.2, focusing on the coordinates of the hull surface's control points. Constraints encompass geometric constraints, wave resistance constraints, and reliability constraints. Geometric constraints mirror those detailed in Sect. 3.5.2, while seakeeping constraints involve seakeeping indexes crucial for gauging the ship's seakeeping performance. Notably, through a comparison of various indexes [8, 9], this example selects the significant wave height pitch angle and significant heave (dimensionless form). Furthermore, reliability constraints govern the failure probability of the seakeeping indexes.

In this instance, the PSO algorithm was utilised for the optimisation process. Concurrently, to mitigate the computational expenses associated with CFD calculations and uncertainty analyses, a Kriging approximation model was chosen to substitute CFD for performance calculations during optimisation. Detailed optimisation settings are provided in Table 6.12.

6.4.2 Optimisation Process

See Fig. 6.9.

The optimisation process unfolded as follows:

(1) Determination of design variables and their corresponding ranges.
(2) Utilisation of the RBF deformation module to generate a new modified hull form based on the outer layer variables, passing it into the uncertainty analysis module.
(3) Utilisation of the LHS method within the uncertainty analysis module to sample the interval variables, namely, wave length and wave height, with sample points at (λ^w, H^w) and $w = 1, 2, \cdots K$, respectively.
(4) Employing the M-H sampling method to sample the strong statistical variable V and the sparse variable T for each group of samples formula, with sample points at (V^η, T^η) and $\eta = 1, 2, \cdots, N$, respectively.
(5) Calculation of the response $\boldsymbol{S}$, representing the average resistance coefficient of the ship in waves and other seakeeping indexes, based on the input uncertain variables and design variables using the established Kriging model.
(6) Computation of the orthogonal polynomial matrix $\boldsymbol{\Gamma}$ based on the input uncertainty, followed by solving the coefficient matrix $\boldsymbol{S}$ of the PCE model under

Table 6.12 Optimisation model

	Reliability-based robust design optimization (RBRDO)
Optimization object	KCS model
Objective	min $\mu(\overline{C}_T)/\sigma(\overline{C}_T)$
Design variables	
X_1	[−0.025, 0.017]
Z_1	[0.146, 0.188]
Y_2	[0.046, 0.075]
Y_3	[0.109, 0.136]
Y_4	[0.066, 0.108]
Y_5	[0.359, 0.393]
Y_6	[0.267, 0.309]
Geometric constraints	
Displacement(∇)	Maximum variation $\pm$ 0.5%
Longitudinal position of center of buoyancy (X_B)	Maximum variation $\pm$ 0.5%
Wetted surface(S)	Maximum variation $\pm$ 0.5%
Seakeeping constraints	
$pitch\ (/\zeta_a)$	$pitch \leq 0.6$
$heave\ (/k\zeta_a)$	$heave \leq 1.4$
Reliablity constraints	
$pitch\ (/\zeta_a)$	$P_f(pitch \leq 0.6) \leq 5\%$
$heave\ (/k\zeta_a)$	$P_f(heave \leq 1.4) \leq 5\%$
Optimizaer	
Algorithm	PSO
Number of particle swarms	60
Number of iterations	20
Approximation model	
Model	Kriging
Number of samples	220

this group of w samples, upon obtaining the response $\boldsymbol{S}$ and the orthogonal polynomial matrix $\boldsymbol{CC}^w$.

(7) Iteration of steps (4) to (6) until all K sample points are processed.

(8) Outputting the sample matrix H and coefficient matrix $\boldsymbol{c}_j$ regarding the interval variable, with specific details provided in Sect. 6.3.

(9) Establishment of the PCE model simultaneously considering interval variables (wave length λ and wave height H), sparse variables (draft T), and strong statistical variables (ship speed V).

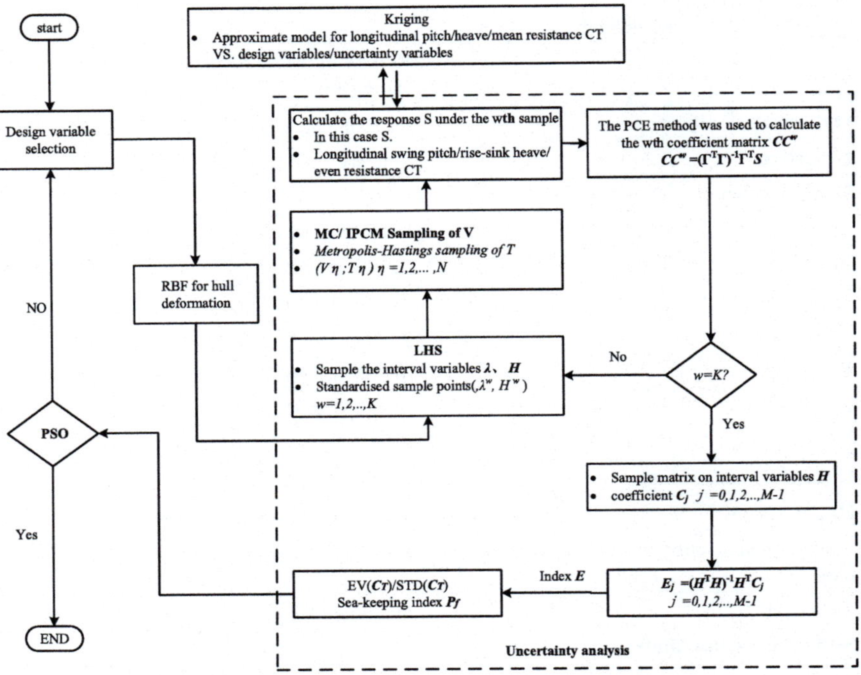

Fig. 6.9 RBRDO process of KCS for wave resistance

(10) Solving the expectation and standard deviation of the objective $\overline{C_T}$, along with the failure probability of seakeeping constraints, based on the coefficients of the PCE model.

(11) Passing the results of step (10) to the optimiser, which assesses whether the optimisation process has reached the predetermined number of iterations. If the set number of iterations is reached, the optimisation concludes; otherwise, steps (2) to (10) are repeated.

6.5 Analysis of Optimisation Results

Simultaneous RBRDO based on aPC and evidence theory were conducted on the standard model KCS. The optimisation results of both methods are presented in Table 6.13.

Regarding the computational costs of the two mixed uncertainty optimisation methods, as indicated in Table 6.13, the count times for evidence theory were $N_c \times N_Z \times N_{XY} = 1 \times 10^3 \times 10^4$ (where N_c represents the number of focal elements; N_Z denotes the number of samples per focal element; and N_{XY} signifies the sampling number of strong statistical variables and sparse variables). Conversely, the count times for the PCE method were $K \times N = 10^3 \times 10^4$ (where K represents the number

Table 6.13 Optimal scheme results of RBRDO

		Initial	RBRDO	
			Evidence theory	PCE
Geometric constraints	$\lvert(\nabla - \nabla_0)/\nabla_0\rvert$	–	0.04%	0.03%
	$\lvert(X_B - X_{B0})/X_{B0}\rvert$	–	0.12%	0.09%
Reliability constraints	$P_f(pitch \leq 0.6)$	10.5%	[0, 4.8%]	4.9%
	$P_f(heave \leq 1.4)$	13.4%	[0, 4.9%]	4.9%
Objective	$\mu(\overline{C}_T)$	12.221	10.957	10.678
	Reduction	–	−10.3%	−12.6%
	$\sigma(\overline{C}_T)$	5.304	3.867	2.242
	Reduction	–	−27.1%	−57.7%
Count times		–	$N_c \times N_Z \times N_{XY}$ $= 1 \times 10^3 \times 10^4$	$K \times N$ $= 10^3 \times 10^4$

of samples of interval variables; N denotes the sampling number of strong statistical variables and sparse variables). In this example, since the evidence space comprising interval variables such as wave length λ and wave height H contained only one focal element, the number of calculations for evidence theory aligned with that of the PCE method. It is evident that an increase in the number of focal elements results in a rapid escalation in the number of calculations for evidence theory, surpassing those of the PCE method. Moreover, in systems solely consisting of strong statistical variables and interval variables, the M-H sampling method could replace the probability distribution point method for sampling strong statistical variables in the PCE method, thereby further reducing the number of samples for strong statistical variables.

Regarding the objective, the principal disparity in solving the objective between the two methods lay in the expression. Owing to the presence of interval variables, the evidence theory-based method yielded the final form not as an exact value but in the form of an interval [7.090, 14.824]. To facilitate comparison with the PCE method, the objective based on evidence theory was expressed in a form akin to the mean and standard deviation through the midpoint and radius of the interval. It is evident from Table 6.13 that while the mean value of $\overline{C}_T$ obtained by the two methods was relatively close, there was a significant disparity in the standard deviation of $\overline{C}_T$. This divergence stemmed from the differing approaches of the two methods in handling interval variables. The intervals of $\overline{C}_T$ and were extensive, resulting in a broader response interval. Consequently, evidence theory represented the interval radius of the objective in a manner akin to the standard deviation, introducing discrepancies compared to the results of the PCE method.

The mean and standard deviation of the $\overline{C}_T$ under RBRDO, obtained through two methods, were smaller than those of the initial scheme. Amidst the influence of uncertain factors, the decrease in the mean of the $\overline{C}_T$ signified that the RBRDO optimal scheme could still achieve optimisation, reducing resistance and enhancing ship performance. Simultaneously, amidst external uncertainties, the reduction in the

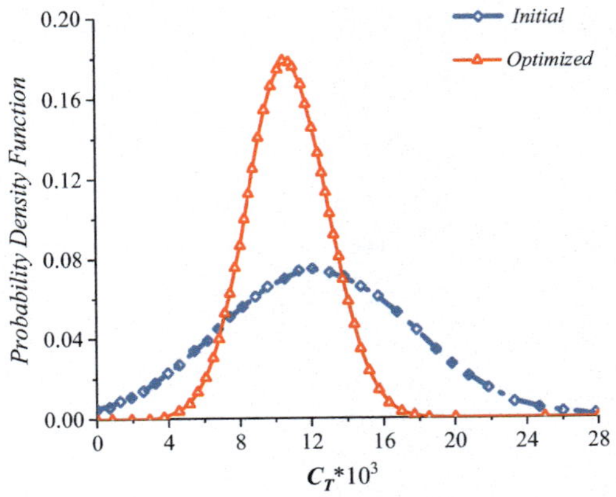

Fig. 6.10 PDF curves under the influence of (λ, H, T, V)

standard deviation of the $\overline{C}_T$ indicated that the optimal scheme displayed improved adaptability to uncertain factors, resulting in fewer performance fluctuations and enhanced robustness while ensuring performance enhancement.

Using the RBRDO optimal scheme obtained through the PCE method as an illustration, under the influence of four uncertain factors (λ, H, T, V), the PDF curve of the optimal scheme is presented in Fig. 6.10. Comparison of the two PDF curves reveals that the standard deviation of the RBRDO optimal scheme was notably smaller than that of the initial scheme. The data points of the $\overline{C}_T$ were concentrated in the high-probability area near the mean, resulting in a slimmer and taller curve. Additionally, the mean value, representing the abscissa at the peak, was smaller than that of the initial plan, indicating a more robust resistance performance amidst uncertainty interference.

Regarding geometric constraints, it could be observed from Table 6.13 that the geometric constraints of the two optimal RBRDO schemes had minimal changes (less than 0.5%) compared with the initial scheme, ensuring the consistency of ship geometry.

Regarding reliability constraints, the failure probabilities of the pitch and heave constraints of the initial scheme were 10.5 and 13.4%, respectively, considering the influence of uncertainty, which exceeded the required 5%. The violation of reliability constraints led to the failure of the scheme. Conversely, for the RBRDO optimal schemes, the failure probability of the pitch and heave constraints obtained by the two methods were lower than that of the initial scheme, significantly enhancing the reliability of the ship type scheme. This ensured that the scheme met the reliability requirements and remained clear of the failure space.

To visually represent the failure probability more effectively, this study plotted the Bel distribution function and Pl distribution function curves for pitch and heave, solved using the evidence theory method, alongside the CDF curves for pitch and heave obtained through the PCE method. As depicted in Fig. 6.11, both the pitch and heave constraints showed failure probabilities of less than 5%, meeting the reliability criteria. Furthermore, Fig. 6.12 illustrates that at the constraint point, the RBRDO

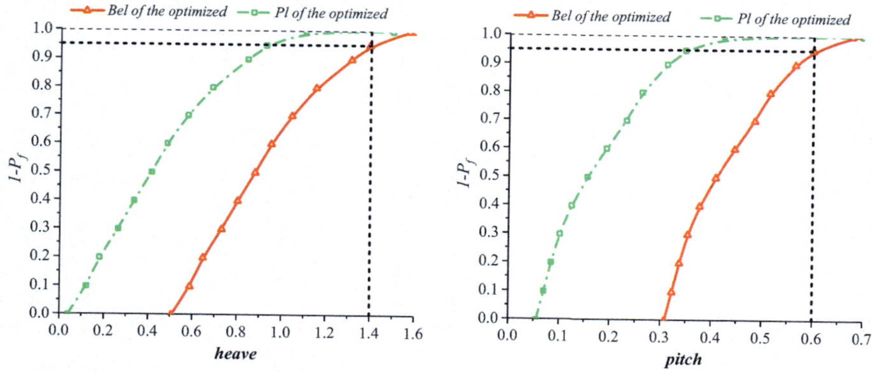

Fig. 6.11 Bel and Pl curves of pitch and heave based on evidence theory

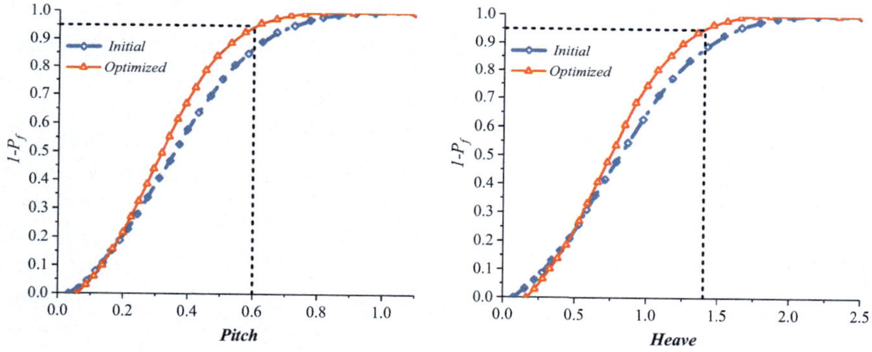

Fig. 6.12 CDF curves of pitch and heave based on PCE

optimal scheme exhibited significantly lower failure probabilities than the initial scheme, meeting the 5% reliability threshold.

Regarding hull lines, taking the optimal hull form derived from the PCE method as an example, it was observed that the hull lines of the RBRDO optimal scheme appeared generally smooth. A comparison with the section and buttock lines of the initial scheme is presented in Figs. 6.13 and 6.14. Contrasted with the initial scheme, the length of the optimal RBRDO plan had increased, aiding in the reduction of pitching and heaving motions. Additionally, the bulbous bow had become larger, higher, longer, and notably upturned. Meanwhile, the bottom of the bulbous bow was slightly thickened, contributing to the reduction of wave-making resistance. The width of the bulbous bow was also increased, and the section lines at the rear of the bow were marginally wider. In summary, the alterations in hull lines were conducive to enhancing ship performance, further validating the optimisation outcomes.

In conclusion, amidst the influence of uncertainty factors, the RBRDO optimal scheme demonstrated superior performance (with a smaller mean of $\overline{C}_T$) and reduced

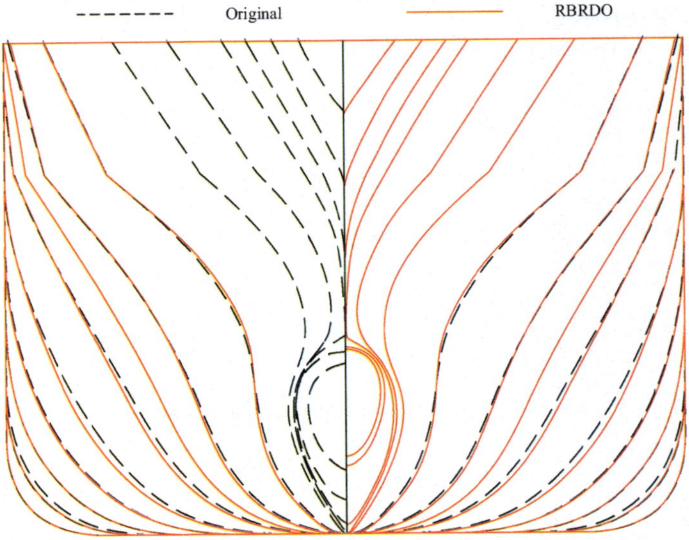

Fig. 6.13 Section lines of RBRDO scheme and initial scheme

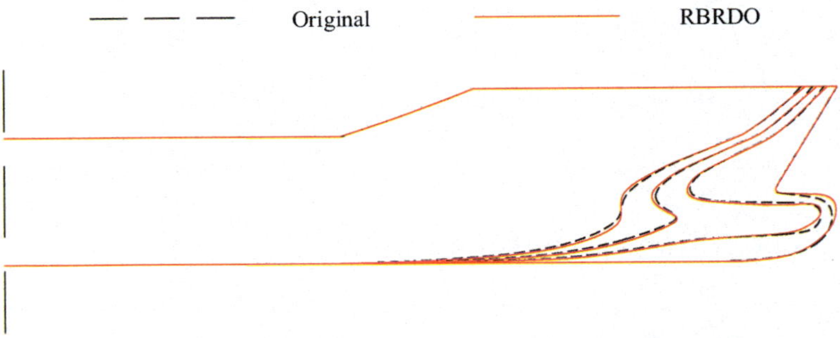

Fig. 6.14 Buttock lines of RBRDO scheme and initial scheme

performance fluctuations (with a smaller standard deviation of $\overline{C_T}$) compared to the initial scheme, while ensuring scheme integrity (with the failure probability of the seakeeping constraints meeting reliability requirements), thus enhancing sailing capabilities.

References

1. Hirt CW, Nichols BD. Volume of fluid (VOF) method for the dynamics of free boundaries[J]. J Comp Phys. 1981; 39(1).
2. Patankar SV, Spalding DB. A calculation procedure for heat, mass and momentum transfer in three-dimensional parabolic flows[J]. Int J Heat Mass Transf. 1972;5(15):1787–806.

3. Tezdogan T, Demirel YK, Kellett P, et al. Full-scale unsteady RANS CFD simulations of ship behaviour and performance in head seas due to slow steaming[J]. Ocean Eng. 2015;97:186–206.
4. Wang S, Su YM, Pang YJ, et al. Numerical study on longitudinal motions of a high-speed planing craft in regular waves[J]. J Harbin Eng Univ. 2014;35(1):45–52.
5. International Towing Tank Conference(ITTC). Proceedings of the 26th ITTC, Practical guidelines for ship CFD applications[C]. 2011.
6. Carrica PM, Wilson RV, Noack RW, et al. Ship motions using single-phase level set with dynamic overset grids[J]. Comput Fluids. 2007;36(9):1415–33.
7. Tezdogan T, Incecik A, Turan O. Full-scale unsteady RANS simulations of vertical ship motions in shallow water[J]. Ocean Eng. 2016;123:131–45.
8. Wu X. Ship manoeuvrability and wave resistance [M]. Beijing: People's Transportation Press; 1988.
9. Feng P, Ma N, Gu X. Research and development of wave resistance forecasting system for ships in real sea area[J]. Mar Eng. 2009;5:6.

Appendix A

A.1 Derivation of the Great Likelihood Estimate of the Distribution to Be Selected

A.1.1 Normal Distribution

Likelihood function

$$L = \prod_{i=1}^{n} \frac{1}{\sqrt{2\pi\sigma^2}} \exp(-\frac{(x_i - \mu)^2}{2\sigma^2}) = (2\pi\sigma^2)^{-n/2} \exp[-\frac{1}{2\sigma^2} \sum_{i=1}^{n} (x_i - \mu)^2] \quad \text{(A1)}$$

logarithmic

$$\ln L = -\frac{n}{2} \ln(2\pi\sigma^2) - \frac{1}{2\sigma^2} \sum_{i=1}^{n} (x_i - \mu)^2 \quad \text{(A2)}$$

induction (math.)

$$\begin{cases} \dfrac{\partial \ln L}{\partial \mu} = \dfrac{1}{\sigma^2} \sum_{i=1}^{n} (x_i - \mu) = 0 \\ \dfrac{\partial \ln L}{\partial \sigma^2} = -\dfrac{n}{2\sigma^2} + \dfrac{1}{4\sigma^2} \sum_{i=1}^{n} (x_i - \mu)^2 = 0 \end{cases} \quad \text{(A3)}$$

Z. Liu et al., *Uncertainty-Based Ship Design Optimization*,
https://doi.org/10.1007/978-981-97-2973-9

Get

$$\begin{cases} \mu^* = \dfrac{1}{n}\sum_{i=1}^{n} x_i \\ \sigma^* = \sqrt{\dfrac{1}{n}\sum_{i=1}^{n}(x_i - \mu)^2} \end{cases} \tag{A4}$$

A.1.2 Uniform Distribution

Likelihood function

$$L = \begin{cases} \left(\dfrac{1}{b-a}\right)^n, \ a \le x_i \le b \\ 0, \ x_i < a\mathrm{orx_i} > \mathrm{b} \end{cases} \tag{A5}$$

The likelihood function is discontinuous and not derivable. If the maximum value L is obtained, then the minimum value b-a is obtained, in which case we have

$$\begin{cases} a^* = \min\{x_i\} \\ b^* = \max\{x_i\} \end{cases} \tag{A6}$$

A.1.3 Exponential Distribution

Likelihood function

$$L = \prod_{i=1}^{n} \lambda \exp(-\lambda x_i) = \lambda^n \exp[-\lambda \sum_{i=1}^{n} x_i] \tag{A7}$$

logarithmic

$$\ln L = n \ln \lambda - \lambda \sum_{i=1}^{n} x_i \tag{A8}$$

induction (math.)

$$\frac{\mathrm{d}\ln L}{\mathrm{d}\mu} = \frac{n}{\lambda} - \sum_{i=1}^{n} x_i = 0 \tag{A9}$$

Get

$$\lambda^* = \frac{n}{\sum_{i=1}^{n} x_i} \tag{A10}$$

A.1.4 Weibull Distribution

Likelihood function

$$\begin{aligned} L &= \left(\frac{\beta}{\alpha}\right)^n \left(\frac{1}{\alpha}\right)^{n(\beta-1)} \left(\prod_{i=1}^{n} x_i\right)^{\beta-1} \exp\left[-\sum_{i=1}^{n} \left(\frac{x_i}{\alpha}\right)^\beta\right] \\ &= \frac{\beta^n}{\alpha^{n\beta}} \left(\prod_{i=1}^{n} x_i\right)^{\beta-1} \exp\left[-\sum_{i=1}^{n} \left(\frac{x_i}{\alpha}\right)^\beta\right] \end{aligned} \tag{A11}$$

logarithmic

$$\ln L = n \ln \beta - n\beta \ln \alpha + (\beta - 1) \ln\left(\prod_{i=1}^{n} x_i\right) - \sum_{i=1}^{n} \left(\frac{x_i}{\alpha}\right)^\beta \tag{A12}$$

induction (math.)

$$\begin{cases} \dfrac{\partial \ln L}{\partial \beta} = \dfrac{n}{\beta} - n \ln \alpha + \sum\limits_{i=1}^{n} \ln x_i - \sum\limits_{i=1}^{n} \left[\left(\dfrac{x_i}{\alpha}\right)^\beta \ln\left(\dfrac{x_i}{\alpha}\right)\right] = 0 \\ \dfrac{\partial \ln L}{\partial \alpha} = -\dfrac{n\beta}{\alpha} + \dfrac{\beta}{\alpha} \sum\limits_{i=1}^{n} \left(\dfrac{x_i}{\alpha}\right)^\beta = 0 \end{cases} \tag{A13}$$

Since the first nth order moments of origin of the Weibull distribution are

$$\frac{\sum_{i=1}^{n} x_i^\beta}{n} = \alpha^\beta \tag{A14}$$

then there is

$$\sum_{i=1}^{n} x_i^{\beta}\alpha^{-\beta}\ln\alpha - n\ln\alpha = 0 \tag{A15}$$

The equation reduces to

$$\begin{cases} \dfrac{n}{\beta} + \sum_{i=1}^{n}\ln x_i - \sum_{i=1}^{n}\left[\left(\dfrac{x_i}{\alpha}\right)^{\beta}\ln x_i\right] = 0 \\ \sum_{i=1}^{n}\left(\dfrac{x_i}{\alpha}\right)^{\beta} - n = 0 \end{cases} \tag{A16}$$

then

$$\begin{cases} \beta\alpha^{-\beta}\sum_{i=1}^{n} x_i^{\beta}\ln x_i - n - \beta\sum_{i=1}^{n}\ln x_i = 0 \\ n - \alpha^{-\beta}\sum_{i=1}^{n} x_i^{\beta} = 0 \end{cases} \tag{A17}$$

For the above equation, solving for $\alpha^{-\beta}$ and bringing it in gives the transcendental function about β.

$$\sum_{i=1}^{n} x_i^{\beta}(n + \beta\sum_{i=1}^{n}\ln x_i) - n\beta\sum_{i=1}^{n} x_i^{\beta}\ln x_i = 0 \tag{A18}$$

Let $\varphi(\beta) = 0$, be solved using Newtonian iteration.

$$\beta_{k+1} = \beta_k - \frac{\varphi(\beta)}{\varphi\prime(\beta)} \tag{A19}$$

A.1.5 Extremal Distribution

Let $\lambda=\frac{1}{\sigma}$.

$$f(x) = \lambda\exp\{-\lambda(x-\mu) - \exp[-\lambda(x-\mu)]\} \tag{A20}$$

likelihood function

$$L = \lambda^n \exp\left\{ \sum_{i=1}^{n} -\lambda(x_i - \mu) - \sum_{i=1}^{n} \exp[-\lambda(x_i - \mu)] \right\} \tag{A21}$$

logarithmic

$$\ln L = n \ln \lambda - \lambda \sum_{i=1}^{n} (x_i - \mu) - \sum_{i=1}^{n} \exp[-\lambda(x_i - \mu)] \tag{A22}$$

induction (math.)

$$\begin{cases} \dfrac{\partial \ln L}{\partial \mu} = n\lambda - \lambda \sum_{i=1}^{n} \exp[-\lambda(x_i - \mu)] = 0 \\ \dfrac{\partial \ln L}{\partial \lambda} = \dfrac{n}{\lambda} - \sum_{i=1}^{n} (x_i - \mu) + \sum_{i=1}^{n} \exp[-\lambda(x_i - \mu)](x_i - \mu) = 0 \end{cases} \tag{A23}$$

The above equation solves for

$$e^{\mu\lambda} = \frac{n}{\sum_{i=1}^{n} \exp(-\lambda x_i)} \tag{A24}$$

Get.

$$e^{\mu\lambda} = \frac{n}{\sum_{i=1}^{n} \exp(-\lambda x_i)} \tag{A25}$$

Substitution and simplification gives the transcendental function on λ.

$$\frac{n}{\lambda} - \frac{1}{n} \sum_{i=1}^{n} x_i + \frac{\sum_{i=1}^{n} x_i \exp(-\lambda x_i)}{\sum_{i=1}^{n} \exp(-\lambda x_i)} = 0 \tag{A26}$$

Let $\varphi(\lambda) = 0$ be solved using Newtonian iteration.

$$\lambda_{k+1} = \lambda_k - \frac{\varphi(\lambda)}{\varphi\prime(\lambda)} \tag{A27}$$

A.1.6 Lognormal Distribution

Likelihood function

$$\begin{aligned} L &= \prod_{i=1}^{n} \frac{1}{\sqrt{2\pi\sigma^2}x_i} \exp(-\frac{(\ln x_i - \mu)^2}{2\sigma^2}) \\ &= (2\pi\sigma^2)^{-n/2}(\prod_{i=1}^{n} x_i)^{-1} \exp[-\frac{1}{2\sigma^2}\sum_{i=1}^{n}(\ln x_i - \mu)^2] \end{aligned} \tag{A28}$$

logarithmic

$$\ln L = -\frac{n}{2}\ln(2\pi\sigma^2) - \ln(\prod_{i=1}^{n} x_i) - \frac{1}{2\sigma^2}\sum_{i=1}^{n}(\ln x_i - \mu)^2 \tag{A29}$$

induction (math.)

$$\begin{cases} \dfrac{\partial \ln L}{\partial \mu} = \dfrac{1}{\sigma^2}\sum\limits_{i=1}^{n}(\ln x_i - \mu)=0 \\ \dfrac{\partial \ln L}{\partial \sigma^2} = -\dfrac{n}{2\sigma^2} + \dfrac{1}{4\sigma^2}\sum\limits_{i=1}^{n}(\ln x_i - \mu)^2=0 \end{cases} \tag{A30}$$

Get

$$\begin{cases} \mu^* = \dfrac{1}{n}\sum\limits_{i=1}^{n}\ln x_i \\ \sigma^* = \sqrt{\dfrac{1}{n}\sum\limits_{i=1}^{n}(\ln x_i - \mu)^2} \end{cases} \tag{A31}$$

A.2 Derivation of Statistical Moments of the Chosen Distribution

A.2.1 Uniform Distribution

First, for the standard normal distribution $x \sim N(0, 1)$, for rth order moments of origin, there are

When r is odd, $E(x^r) = 0$;
When r is even.

$$E(x^r) = \frac{1}{\sqrt{2\pi}} \int_{-\infty}^{+\infty} x^{r-1} \mathrm{d}(-e^{-\frac{x^2}{2}}) = (r-1)\frac{1}{\sqrt{2\pi}} \int_{-\infty}^{+\infty} x^{r-2} \mathrm{d}(-e^{-\frac{x^2}{2}})$$
$$= (r-1)E(x^{r-2}) = (r-1)(r-3)\cdots 3 \cdot E(x^2)$$
$$= (r-1)(r-3)\cdots 3 \cdot 1 \tag{A32}$$

For a general normal distribution, $Y \sim N(\mu, \sigma^2)$, there are

$$E(Y^k) = E\left[(\mu + \sigma \cdot x)^k\right] = E(\sum_{r=0}^{k} C_k^r \sigma^r x^r \mu^{k-r}) = \sum_{r=0}^{k} C_k^r \sigma^r \mu^{k-r} E(x^r) \tag{A33}$$

where $C_k^r = \frac{k!}{r!(k-r)!}$。

A.2.2 Uniform Distribution

For a uniform distribution $[a, b]$, there are

$$E(x^r) = \int_a^b \frac{1}{b-a} x^r \mathrm{d}x = \frac{1}{b-a}\frac{x^{r+1}}{r+1}\Big|_a^b = \frac{1}{b-a}\frac{b^{r+1} - a^{r+1}}{r+1} \tag{A34}$$

A.2.3 Exponential Distribution

For the exponential distribution, there are

$$E(x) = \int_0^{+\infty} x\lambda e^{-\lambda x} \mathrm{d}x = \frac{1}{\lambda} \tag{A35}$$

$$E(x^r) = \int_0^{+\infty} x^r \lambda e^{-\lambda x} \mathrm{d}x = -\int_0^{+\infty} x^r \mathrm{d}(e^{-\lambda x}) = \frac{k}{\lambda} \int_0^{+\infty} x^{r-1} \lambda e^{-\lambda x} \mathrm{d}x$$
$$= \frac{k}{\lambda} E(x^{r-1}) = \frac{k}{\lambda}\frac{k-1}{\lambda} \cdots E(x) = \frac{k}{\lambda}\frac{k-1}{\lambda} \cdots \frac{1}{\lambda} = \frac{k!}{\lambda^r} \tag{A36}$$

A.2.4 Weibull Distribution

$$E(x^r) = \int_0^{+\infty} x^r f(x)\mathrm{d}x = \int_0^{+\infty} x^r \frac{\beta}{\alpha}\left(\frac{x}{\alpha}\right)^{\beta-1} \exp\left[-\left(\frac{x}{\alpha}\right)^{\beta}\right]\mathrm{d}x \quad \text{(A37)}$$

Let $w = \frac{x}{\alpha}$, there are

$$E(x^r) = \int_0^{+\infty} \alpha^r w^r \frac{\beta}{\alpha} w^{\beta-1} \exp[-w^{\beta}]\alpha \mathrm{d}w = \alpha^r \beta \int_0^{+\infty} w^r w^{\beta-1} \exp[-w^{\beta}]\mathrm{d}w \quad \text{(A38)}$$

Let $z = w^{\beta}$, than $w = z^{1/\beta}$, $\mathrm{d}w = \frac{1}{\beta} z^{\frac{1}{\beta}-1}\mathrm{d}z$, there are

$$E(x^r) = \alpha^r \beta \int_0^{+\infty} z^{\frac{r+\beta-1}{\beta}} \exp(-z) \frac{1}{\beta} z^{\frac{1}{\beta}-1}\mathrm{d}z = \alpha^r \int_0^{+\infty} z^{\frac{r}{\beta}} \exp(-z)\mathrm{d}z = \alpha^r \Gamma(\frac{r}{\beta}+1) \quad \text{(A39)}$$

in Eq. (5.28) Γ represents the gamma function.

A.2.5 Extremal Distribution

For the extreme value distribution, let $z = \frac{x-\mu}{\sigma}$, there are

$$E(z^r) = \int_0^{+\infty} z^r \exp(-z) \exp(-e^{-z})\mathrm{d}z \quad \text{(A40)}$$

Let $y = e^{-z}$, than $z = -\ln y$, there are

$$E(z^r) = \int_0^{+\infty} (-\ln y)^r y \exp(-y)\mathrm{d}(-\ln y) = (-1)^{r+1} \int_0^{+\infty} (\ln y)^r \exp(-y)\mathrm{d}y \quad \text{(A41)}$$

(5.30) needs to be solved numerically.

A.2.6 Lognormal Distribution

For the lognormal distribution $x \sim LN(\mu, \sigma^2)$, there are

$$E(x^r) = \int_0^{+\infty} x^r \frac{1}{\sqrt{2\pi\sigma^2}x} e^{-\frac{(\ln x-\mu)^2}{2\sigma^2}} \mathrm{d}x \tag{A42}$$

Let $y = \ln x$, there are

$$E(x^r) = \int_0^{+\infty} e^{yr} \frac{1}{\sqrt{2\pi\sigma^2}} e^{-\frac{(y-\mu)^2}{2\sigma^2}} \mathrm{d}y = e^{\mu r+\frac{1}{2}r^2\sigma^2} \int_0^{+\infty} e^{yr} \frac{1}{\sqrt{2\pi\sigma^2}} e^{-\frac{\left[y-(\mu+r\sigma^2)\right]^2}{2\sigma^2}} \mathrm{d}y = e^{\mu r+\frac{1}{2}r^2\sigma^2} \tag{A43}$$

Appendix B

See Tables B1, B2 and B3.

Table B1 Design variables

Design variable	Range
Length L/m	$150 \leq L \leq 274.32$
Breadth B/m	$20 \leq B \leq 32.31$
Depth D/m	$14 \leq V_k \leq 20$
Draft T/m	$10 \leq T \leq 11.71$
Block coefficient C_B	$0.63 \leq C_B \leq 0.75$
Design speed /kn	$14 \leq V_k \leq 20$

Table B2 Constraints

Constraints	Range
Length width ratio	$L/B \geq 6$
Length depth Ratio	$L/D \leq 15$
Length draft ratio	$L/T \leq 19$
Froude number	$F_n \leq 0.32$
Deadweight capacity	$25000 \leq DW \leq 500000$
Practical constraints on draft and deadweight	$T - 0.45DW^{0.31} \leq 0$
Practical constraints on draft and depth	$T - 0.7D - 0.7 \leq 0$
Practical constraints on stability	$0.07B - 0.35T - \frac{(0.085C_B - 0.002)B^2}{TC_B} + 1 + 0.25D \leq 0$

Z. Liu et al., *Uncertainty-Based Ship Design Optimization*,
https://doi.org/10.1007/978-981-97-2973-9

Table B3 Calculation model

Name	Formula
Steel weight	$W_S = 0.034L^{1.7}B^{0.7}D^{0.4}C_B^{0.5}$
Outfitting weight	$W_O = L^{0.8}B^{0.6}D^{0.3}C_B^{0.1}$
Mechanical and electrical weight	$W_M = 0.17P^{0.9}$
Bare weight	$LS = W_s + W_o + W_m$
Power coefficient a	$a = 4977.06C_B^2 - 8105.61C_B + 4456.51$
Power coefficient b	$b = -10847.2C_B^2 + 12817C_B - 6960.32$
Froude number	$Fr = \frac{0.5144V_k}{\sqrt{9.8065L}}$
Displacement	$\Delta = 1.025LBTC_B$
Power	$P = \frac{\Delta^{2/3} V_k^3}{a+b\times Fn}$
Deadweight capacity	$DW = \Delta - LS$
Daily fuel consumption	$DC = 0.2 + \frac{0.19\times 24}{1000}P$
Fuel carrying capacity	$FC = DC(D_S + 5)$
Crew, shops and water	$CSW = 2DW^{0.5}$
Deadweight of cargo	$DW_C = DW - FC - CSW$
Days in transit	$D_S = \frac{5000}{24V_k}$
Days in port	$D_P = 2\frac{DW_C}{8000} + 1$
Annual round trip	$RTPA = \frac{350}{D_S+D_P}$
Cost	$C_S = 1.3\left(2000W_S^{0.85} + 3500W_O + 2400P^{0.8}\right)$
Capital	$C = 0.2C_S$
Running cost	$C_R = 40000DW^{0.3}$
Voyage cost	$C_V = 1.05DC \times D_S \times 100 + 6.3DW^{0.8}$
Annual cost	$C_A = C + C_R + C_V \times RTPA$
Annual goods	$AC = DW_C \times RTPA$
Freight charge *TC* (£/yr)	$TC = C_A/AC$

The manufacturer's authorised representative in the EU is Springer Nature Customer Service Centre GmbH, Europaplatz 3, 69115 Heidelberg, Germany. If you have any concerns regarding our products, please contact ProductSafety@springernature.com

Printed and bound by CPI Group (UK) Ltd, Croydon, CR0 4YY
01/10/2026
02225919-0012